AF801705

Awkward news for Greenies

and everyone else

Awkward news for Greenies

and everyone else

Dr J. Mark Dangerfield
Ashley Bland

First Edition

Think Truth – Sydney, Australia

First published April 2009

Copyright © John Mark Dangerfield & Ashley Bland

All rights reserved.

No part of this publication may be reproduced, stored in a retrieval system or transmitted in any form by any means, electronic, mechanical, photocopying, recording or otherwise, without prior written permission of the authors.

Dangerfield, J. Mark, 1961-
Awkward news for Greenies and everyone else
Includes index.
ISBN 978-0-9806360-0-0
1. Environmentalism. 2. Ethics. 3. Ecology. 4. Land, energy & natural resources.
I. Bland, Ashley
II. Title

Cover design by Mark Dangerfield and Alyssa Naccarella

Think Truth
PO Box 4240
Winmalee
NSW 2777
Australia

www.thinktruth.com.au

10 9 8 7 6 5 4 3 2 1

Table of contents

Part Three - Collective

Part Four - The awkward news for Greenies

Preface

"Think!" the old man said as the young boy looked to the floor. "Just think about it."

It was hard sometimes, as a child, to understand what our elders meant when they told us to think. However, we soon learnt that trying our best was a much better option than a kick up the bum, so we'd muse, cogitate and hope to do something different next time. As we grew up we gained a sense of what thinking could do for us, apart from staying out of trouble. We would think about sex, love, money, the weather, a hamburger. Soon enough thinking became our natural state. Now, as adults, we think all the time.

This is not all bad. Human thought has delivered the most amazing ideas, things and technologies. It was thinking that took us to the moon and led to treatments for cancer. We have thought up ways to build amazing shelters, grow hundreds of food crops, mine for useful minerals and divert water to our benefit. It was also thinking that led to the development of nuclear weapons and third world sweatshops - so we know that being smart does not always lead to positive outcomes for everyone.

Now, as we begin the 21st century, an awkward topic has begun to demand, even scream out for our attention. There is a problem with the Environment. Climate change, global warming, toxic pollution, poor health, animal and plant extinctions, drought, water supply and quality, and any number of related issues seem to need thinking about. A group within society promotes these issues with a passion. Environmentalists and greenies say that the writing on the wall has us all racing toward impending doom. There is truth behind what they say. The issues are real. But is our response to the issues working? Given that we are spectacularly good at thinking about problems and solving them, why haven't we already fixed this? Why does it seem so hard?

This book deals with how we think about the environment, the awkwardness of humanity in the 21st century. How do we understand our predicament and how might we include the unseen challenges not far ahead in the race? It is about where the real problems lie.

When we are sick we go to the Doctor and expect advice, treatment or a tablet that will fix us. Increasingly though we are told that lifestyle, food, exercise, stress levels and our general wellbeing all play an important role in our health. This means that, even from the medic, there is not one single magic solution to our illness because there is rarely one single cause or problem. Yet this is how we search for solutions. We hope that our thinking will deliver *the* solution. It would be nice if there were a simple solution to how we will provide shelter, food and comfort to 6.5 billion souls. As though it was something we could medicate.

Currently, for example, we are being told that if we reduce our carbon emissions the problems of bad weather and rising seas will go away. We are told that if we lock up enough reserves our animals will survive and our landscapes will be healthy. The implication from the news-grabs is that if we can just fix this problem, then all will be well. This book explains why it won't.

This book is about thinking a little differently. A thinking that includes feeling and making choices. It is about finding the thinking, feeling and choosing about the environment that may shift our current trajectory. We will show that this sort of process is complicated and sometimes painful, but we all know one thing. If we do in the future as we did in the past, yet expect a different outcome, we are mad.

Information is to thinking what bricks are to building. No amount of thinking with poor information will lead to optimal answers. So we have constructed this book around discussions of facts and figures, ecological concepts and a few moral dilemmas thrown in for good measure. In paring down the complexity we have settled on three main sections: Part One - Users, Part Two - Touch the earth and Part Three - Collective.

We have also taken three different approaches to generating thought. Each part begins with an overview; an exposition of the issues. Then we provide more detailed essays on the concept; the salient facts. The third approach is to find the feeling behind the facts through fictional stories. We hope this mixture will reach people with different ways of thinking and we expect some styles will appeal to you more than others.

In Part Four – The awkward news for Greenies, we discuss where the problem lies.

We have tried to keep jargon and technical terms to a minimum and where we have found no other way we have provided footnotes.

Somewhat frustratingly for the sales department, this book may make you feel uncomfortable at least once. In fact, we hope that it does, because there are times when we benefit from discomfort. But do not do anything rash, for such feelings often lead to a deeper understanding, inspiration and wisdom. We will get through this. We just need to have a better think about it.

About us

The world population clocks tell us that on 15th November 2007 there were 6,630,265,436 humans on the earth; give or take. This is more than three times the number alive on the eve of the Second World War and double the number alive just 40 years ago. In downtown Sydney, Australia, two of those 6.6 billion folk met for the first time. At 3pm on that sunny day **Dr J. Mark Dangerfield**, a bald scientist, and **Ashley Bland**, a lanky dude with just a hint of green, shook hands.

Mark had trained as an ecologist in the mud and rain of the UK, but escaped, first to a decade of solace in the warmth of African savannas and then, in 1996, to the forests and paddocks of Australia. After completing a Post-doctoral Fellowship at the University of Zimbabwe he held faculty positions at the University of Botswana, Macquarie University and University of Technology Sydney helping undergraduate and postgraduate students to understand biodiversity, ecology, statistics, soil biology and wildlife management, and in 1999 he received an Outstanding Teacher Award from Macquarie University. His research into the ecology of soil animals and the measurement of biodiversity attracted over $1.5 million in competitive funding and he has published more than 120 research papers and reports. In 2001 he took a break from academia as a founder and CEO of Biotrack Australia Pty Ltd, a company that developed and provided services to measure the health and performance of the environment. Throughout Mark served on many government committees and is the science representative on the NSW Natural Resource Advisory Council. At the time of this first meeting with Ashley he was Associate Professor and Director of the Centre of Expertise for Environment at the University of Technology Sydney.

Ashley grew up on a farm, catching tadpoles, planting broccoli and learning in a very real sense the connectedness of all life and the ultimate dependence of humanity on a healthy environment. After studying Environmental Engineering his desire to experience outback Australia took him to far north west Queensland where, with long-hair (but no beard) he took on the formidable role as the mine's first Environmental Officer, working with the Kalkadoon Aboriginal peoples and convincing miners that they should change their ways. This was long before "environment" was the buzzword it is now. Ashley returned to the farm to write and record original music, living a double life of organic farmer and dynamic performer that took him from Sydney to Kalgoorlie to New York, and continues today. His passion for environmental action brought him back off the farm into a career with Conservation Volunteers Australia where he still works, this time convincing companies and their staff to become involved in protecting our natural resources. In-between all that Ashley spent 3 years overseas, traveling and working as a photographer and videographer on a cruise ship learning skills which have led to he and his wife setting up an independent production company.

Knowledge of these two varied but different resumes would not tell you how things would go. As with all first encounters people are cautious.

The meeting was to discuss solutions to how volunteers might monitor the conservation outcomes of environmental restoration works. It was an intriguing challenge. In a lively discussion that went over its allotted time, ideas converged on an solution where some science savvy, helped along by some information technology, could really help get over the problem of not having enough information on the environment. Why not engage anyone with an interest, a little training and some dedication to generate useful environmental information. Put information gathering into the hands of those doing the work on the ground, the farmers, foresters, conservation groups, and even the school kids. This was an exciting idea.

Unfortunately it was not an easy concept to sell. The conventional wisdom says that only scientists and those trained in the sciences are the generators and rightful custodians of environmental information. How could anyone else provide information with sufficient credibility?

In the months that followed Mark and Ashley had several more conversations, usually accompanied by huge pots of tea. The information gathering idea ran its course and hope lingers that funds might be secured to get the concept moving.

Later, the discussions broadened to humans and the world they inhabit. The result was *think truth environment.* A place where thoughts on the environment are not rhetoric, spin or agenda driven but try to achieve the truth, at least the truth as we understand it.

This book is the first publication from this collaboration.

www.thinktruth.com.au

Part One

Users

Chapter 1

Overview

Users

What all living things do

The Pied Currawong is a shoe-box sized Australian native bird; mostly black with a little white, a long sharp beak and a distinctive call. Currawongs eat just about anything including fruit, insects and small vertebrates like lizards or chicks in the nest. In native bushland you will find occasional birds or small groups of Currawongs nesting together. In orcharding districts, such as around Bathurst in Central New South Wales, Australia, large flocks gather and annoy the farmers by eating fruit. For Currawongs times are good in orchards and the population grows.

Currawongs also eat the chicks of small birds and with many of them around the small bird population inevitably decreases. The farmers invent an effective way of excluding these annoying creatures by netting the fruit trees. The hungry Currawongs extract a heavy toll on the small bird population that becomes a major food source rather than an opportunistic feed. Yet it is not enough. Over winter many of the now lean Currawongs die in the cold. For those that survive through to spring there is no access to fruit and very few small birds, so it becomes a struggle to raise their own chicks. But this is good news for the surviving small birds that can now raise their young with little risk of Currawongs eating them. And on it goes.

There is a twist in this tale. The farmer is happier now that he has protected his resource, the fruit. Except that without a healthy population of small birds eating the caterpillars of the cherry moth, the cherry slugs, as they are known locally, increase in number and become a problem. The innovation master thinks again and decides that an application of insecticide spray will control the slugs. Unfortunately the spray also kills may other insects and slows the recovery of the small bird populations.

This simple example shows how interdependent the 'web of life' is. A change in one population will affect many others.

The universal pattern in nature is that every organism is driven to create more of itself, to expand and make more. First an organism ensures that it can survive long enough. More-making requires time to acquire and convert resources (food, nutrients, water, sunlight) into cells and tissue that can be used to create. Next an organism will grow and mature, storing resources for the big effort until finally, it will make more of itself through reproduction. These three fundamentals, *survival*, *growth* and *reproduction,* define life.

Imagine a forest with flowering plants that produce nectar. Birds will eat the nectar as a source of energy to fuel their more-making activity. There are also insects that survive on this nectar. The production of nectar is finite, determined by the more-making of the plants, constrained by the environmental conditions. If there are a lot of insects then the birds have little to eat and may die out. Too many birds and the same will happen to the insects. It is never quite this simplistic but if we imagine for a moment that birds and insects are restricted to nectar, it is easy to see how excess appropriation by one affects the other. Most organisms live within this fluctuating balance. They adjust their more-making to take advantage of the good times when nectar is plentiful and slow down when it is scarce. They do not seek to increase the nectar supply.

A similar pattern appears in the relationship between predator and prey, that is, the eater and the eaten. When there is plenty of prey (food), times are good for the predator which will therefore breed more successfully, increase in number and so eat even more of the food source. Over time the increase in population levels-out or slows down because now there are so many predators that they get a smaller and smaller percentage of the food source and life is not as easy for them.

The feature that determines these cycles, and drives some of the innovation, is that there is a limited capacity of the Earth to generate resources (food). In the orchard example above there is a finite number of trees that can fit into a paddock before they start to be limited by competition between themselves for water and nutrients – even with the farmer adding more. There are only so many cherries that a tree can grow, determined by the ability of the tree to convert energy and nutrients into fruit. Organisms innovate to find ways around resource shortages but ultimately the environment constrains more-making.

The amount of plant life that can grow on a given area of land we call terrestrial primary production. It is dependent on the amount of water, nutrients and sunshine available, with more of all three generally resulting in more life. Think of deserts compared to rainforests for example. In wilderness areas, the places where the influence of humans is reduced, the vegetation grows according to these basic factors (water, nutrients and light) and other organisms evolve to use the plants as a resource for their own more-making (secondary production). Amongst the secondary producers the multiple interacting predator prey cycles mean that no single species can do too well and therefore take out too much of its food source. There is a kind of equilibrium established, more strictly it is a multitude of checks and balances brought about by the consequences of more-making, and the same rules apply everywhere on Earth.

What humans do

Life is amazingly inventive. From virus' adapting to get around the latest antibiotics to crows bending wire to access food, life is constantly inventing ways to solve problems. Life also solves the same problem in a myriad of ways. A glance in the florist's window or a stroll through a meadow in spring shows flowers in profusion. No two are the same. Each flowering plant species has a unique flower, even though this variety is all to solve one essential problem of producing pollen and getting it to a receptive stigma on another plant of its kind.

Animals, with their mobility, speed and dexterity, are the masters of invention because they can add behaviours to the normal mix of chemical, physiological and reproductive novelty. This can even include a sharing out of tasks. For example, Grey meerkats (suricates) organize themselves so that one of their group is always on the look-out for predators, dingoes hunt in packs to save energy and increase the likelihood of eating, and Apostle birds nest in groups to share the burden of raising young.

Arguably the bundle of cells known as *Homo sapiens* is the most successful problem solver there has ever been. In fact, we love it. In the 200,000 years of our existence we have solved the problems of food, shelter and safety over and over, in a million different ways. Indeed, so good are our skills at problem solving that we seem to invent new problems just to solve them. Anyone who has opened their pre-Christmas junk mail will be astounded to learn of solutions to a whole range of problems they never knew they had.

Humans have become so good at innovation that they have broken the rules. Predators do not limit us anymore. Our shelters are so effective that we no longer have to fear big cats or snakes or the vagaries of the weather. We have invented drugs to cure us when we are affected by pathogens and thrown pesticides at the transmitters of disease. We have also learnt how to change the primary production to suit ourselves. Our appropriation of terrestrial primary production is so effective that multiple other species cannot survive[1] and in some places we have pushed the system to collapse.

Up until a few thousand years ago the primary problem for *Homo sapiens* was how to survive. How do I get enough food? Where can I live safe and warm and protected from the elements? How do I give birth and survive to nurture the child? How do I survive this skirmish with my neighbors, the Neanderthals, or avoid that large cat? And although it might not feel so stark, we still face versions of these problems today.

---

[1] The modification of habitat to appropriate primary production is the most significant cause of biodiversity loss.

It is most obvious in the known flashpoints around the world but also among the 2 billion of us that live on less than US$10 per day. Survival problems are even perceived by the more fortunate of us as we seek to pay the bills in the daily grind of trading time for money. And yet our overwhelming desires to persist, grow and create are as strong as ever. In this innate drive to make more we are at one with all other organisms. However, our ability to increase the odds of success has no peer.

We have also mastered the innovation of 'safety in numbers', and extended it. Humans have taken working together to extremes and proven to be amazingly adaptable. Even before we developed basic technology or farming practices around 10,000 years ago, we had formed small groups organized enough to see us go to all ends of the earth from ice fields to deserts.

All this was against the odds really. We have no natural weaponry, no bite or sting. We cannot fight very well. It takes us a biological age to have babies and then it's only one or two, sometimes killing the mother in the process. By the standards of other mammals of our size, we cannot move very fast, our skin is soft and our senses pretty average. Yet, in terms of more-making, we have been spectacularly successful.

It was our ability to think, communicate, work together and adapt our environment in inventive ways that overcame these weaknesses. We managed to control fire, construct shelters, make tools, and invent useful new things like rope, the wheel and, eventually, agriculture. More than just overcome weakness, this ability to invent has led to great success. The result is a current human population approaching 7 billion that is spread across the entire planet, living mostly in groups, some of up to 15 million individuals. Our ability to invent is so developed that there seems no limit to our innovation and no problem that we cannot solve. No limit in our ability to find ways of making more.

Let's dwell on that little phrase "adapt our environment" for a moment. It is the one notable difference between humanity and the rest of nature. All organisms make more, but they do it for the most part in response to what the environment has to offer. Human beings have become very skilled at changing our immediate surroundings in order to make our lives easier. We change the environment to suit ourselves, to satisfy that innate desire by making survival, growth and reproduction more secure.

The common bracken fern *Pteridium aquilinum* is a spectacularly successful plant. Like humans it has managed to find a home on every continent, except Antarctica, and in all environments except for hot and cold deserts. It does this by producing millions of microscopic spores, able to be carried vast distances on the wind and each capable of starting a new plant. Bracken ferns are not that fussy about where to live. If a spore lands where conditions are suitable, it grows. If conditions don't suit there is nothing the tiny spore can do about it and it fails. The difference between bracken and us is that bracken fits into the existing environment whereas we change it to suit us.

When humans arrive they will most often immediately start cutting plants down to create shelter, burn some of the wood for heat, hunt everything in sight and burn more vegetation to cook the catch. Before long nature has been bent to our needs.

Increasingly we have bent nature to the extreme. Our unique ability is to change what is around us to make it easier to live. We do it all the time. Indeed we have enshrined this ability to bend nature in our cultures and in holy texts as a God-given right. We change the world around us without thinking. It is what we do, and it is different to everything else.

We are the ultimate users.

What it means to be users

By the time the Neanderthal lineage died out around 25,000 years ago, *Homo sapiens* had migrated across the planet. We were everywhere, although not long before this we may have been close to extinction. Some paleontologists have used genetic evidence to suggest that there may have only been a few thousand *H. sapiens* alive 70,000 years ago, just before the big migration. Then something unique happened. Whatever innovation we found, it worked. We extended our range to live across the world albeit in small groups and at low densities.

In these early times, and for the majority of our existence as a species, we were just another one of the many in the mix. There were perhaps only one million people when agriculture was invented 10,000 years ago. Even this amazing solution to the food problem took a while to have an effect on human numbers. In biblical times there were perhaps 200 million people and not until the 1800's did we hit one billion (1,000,000,000).

Two hundred years later, a blink in our history, and it is all very different indeed.

An hour from now approximately 8,000 new human souls will have been added to the human population of 6,756,500,000. In a week there will be 1,344,000 more births than deaths.[2] Our success in more-making has become staggering. Not only have the shackles been broken, they have been smitten into the dust.

We have even found ways for people whom for biological reasons, or reasons of sexual orientation, cannot have a child to obtain one through fertility treatments or surrogacy. We don't like kids when they poop, whine or pierce their ears but we want them with a passion.

Each one of us still has that primal urge to make more. Not all of us may desire bucket loads of offspring but we all seem to want one, or maybe a couple of them. It is what we do.

Our desire to make more extends beyond a little bundle of joy to call our own. It is well established that all of us, regardless of income or socio-economic position, respond to wealth questionnaires by saying we would be "that much better off" with just 10% more money. A little bit more, just to be comfortable, just to be safe.

So along with our increase in number we have an increase in need. Not only are we appropriating the Earths primary and secondary production - the resources we need to stay alive and reproduce - we are also using them well beyond having those basic needs met, just to feel better.

It could be worse. Fortunately we have invented a way of making more that is decoupled from biological more-making. We call it making money. This is really quite clever. There is

---

[2] Global human population growth rate is at 0.5%, lower than the peak of 2% in the 1960's, but because of overall population size this percentage translates to 80 million new people per annum.

much to be said about this aspect of what we do and yet we cannot eat money. Even the very wealthy require food and water.

As children we learn at some point that the cookie jar is empty. That's OK, because Mum or Dad will go out and buy another packet. We have a sense that the resources we dig out of the ground, like coal and steel, are finite and one day there will be no more to mine. But for food we expect nature to keep on delivering, year after year so that there is always another packet of cookies in the shop to end up in our jar. But what if we are mining the ability of nature to keep producing for us? What if our appropriation of the Earth's terrestrial primary production is so great that, at the same time as we increase in number, all wanting more all the time, nature's ability to provide is diminishing?

Appropriation that leads to a crash is not a problem for nature or the environment. Soon enough a new equilibrium is established, albeit initially as a simplified system with fewer species (reduced biodiversity). However, simpler systems are less stable; prone to massive fluctuations in production and in predator prey cycles. Examples of this abound from blue-green algae outbreaks in our dams and rivers that make water undrinkable, to locust plagues destroying crops.[3]

Clearly, if we are concerned about the capacity of nature to provide us with the resources we need to survive, then it would be sensible to limit the amount of primary production that we use for ourselves. We might invest our innovation and problem solving skills in limiting our resource use. Perhaps we would start with 'wants' before we even talked about reducing 'needs'. The difficulty is that we are now talking about human nature, belief systems, politics and the innumerable vagaries of multiple and diverse societies. Even if we acknowledge a need to reduce our consumption, for all our own sakes, we do not know by how much. We also do not have a system for sharing out the reduction among ourselves. We have danced around this sort of question for some time. The problem is that we struggle to supress our fundamental desire to make more.

---

[3] Jarrod Diamond's book *Collapse* catalogues examples of the impact of failing ecosystems on human civilizations.

There are essentially three choices. Kill off some people. Hmmm, this is problematic but happening by complicit neglect anyway. Redistribute the same amount more equally amongst and within Nations. Hmmm, we use less so they can use more? Don't like the sound of that. Or, for all the gluttons on the planet... use less. Ummm, but I can't! It hurts.

The choice hurts because more is innate. Each year drug and alcohol addiction costs millions of dollars to the Australian taxpayer, both in the ugly consequences of abuse and in trying to rehabilitate users. This kind of addiction we know about, although we are not good at understanding it. Addicts are shunned for the most part and the collective problem we leave to underfunded government departments and charities. But we too are addicts; our use of resources has become an addiction, an addiction to making more. So far, however, this weakness does not hold a stigma.

As we will explore later in this book, in the western world our lives have become so removed from the cycles of nature that it is easy to miss the direct link between our consumption of resources and a change in nature. A look around with eyes open shows us that change is occurring.

"Not a problem" many will say because we have technology. We have more than enough innovation and engineering skill to solve these problems. There is no doubt that technology will help, however, we must be careful that we are not simply finding another way to feed the addiction or just a way to treat the symptoms. Take, for example, water purification. We can filter out the toxins produced by blue-green algae outbreaks but this does nothing to limit our use of water, nor does it address the issues in the catchment that are causing the nutrient imbalance.

One step in any addiction scenario is to try to cut back. We know it's still harmful but we hope that in using less we minimize the damage. Technology requires energy and materials so the technology solution tends to help us use more.

A further impediment to cutting back is that restraint is not supported by capitalism. In fact, every business, at its core, embraces more-making because, as a rule, more stuff equals more money. And we all depend on this 'more money' to feel just that little bit better.

There may be some external fixes, but for users to really 'use less' they must make a moral choice and find a balance. This is a tough gig; hard at an individual level and hard at a societal level.

There are bookshelves full of tomes dedicated to the "best" way to organize society to deal with these dilemmas and it's not the intention to summarise them here. What we need to understand is that two incredibly complicated webs are interacting: the web of life and the web of commerce. Pull on any particular strand and the entire network is affected to varying degrees. But, unless you are dead, you can't help pulling on these strands. It is possible to understand when and how we pull on the webs of nature and commerce and to see the consequences.

The way to do it is to make an informed choice.

Knowledge and understanding, especially of the complexity and the consequences of our actions, is available to us. All we need to do is become aware of it. Then our choices can be made with clarity. To acknowledge that we are users is both a fundamental and a powerful admission. This is the first step to understanding the addiction that is having an impact on the rest of life on Earth.

We will still have to use 'stuff' and, in fact, our society will fall over if we don't. It is enough to know and to become aware of the fact that everything you use has a consequence. It is possible to feed the addiction but without the bad health consequences.

Ultimately humans are incredibly efficient resource users, but, for the most part, we are ignorant of the consequences. In this section an essay and three stories explore this concept of human beings as users of the environment.

Chapter 2

Essay

Dodging the grid

In western countries houses have roofs that rarely leak, running water, a sewage system, a place to cook, cozy bedrooms and a living area where it is pleasant to sit and relax. Most homes will also have a refrigerator, hot water in the bathroom, a telephone and a television. There will be any number of game consoles, mobile phones, computers, sound systems, and kitchen appliances that bleep, blurt and bounce sound around the room.

If our great-grandparents had taken a gander into the future, the profusion of technology in a modern home would be unfathomable to them. They would still be getting over carriages that moved without horses and a talking box with Bakelite knobs. Today we tend to take our comfort and technology for granted, yet how often have you had the conversation that includes: "what did we do before mobiles?" Somewhere deep down we know that the advance of gizmos and gadgets to make life easier is both staggering and something that we should be grateful for; and we also know that technology is with us and, baring catastrophe, we are unlikely to go back to washboards and candles.

We tread along a technology path that stretches out ahead of us, meandering inevitably to a point in the far distance. We can only see a little way forward along this path to touch screen mobile phones, hybrid cars and gigabyte downloads. The middle distance is a little hazy but we can just make out bionic body parts, artificial intelligence in robot home helps, and abundant fusion energy. Look further ahead we can only guess at the path that our grandchildren will tread.

Further still and we have as much trouble imagining the technology future for our great-grandchildren as our ancestors would have done thinking about our lifestyles.

Most of us just trust, however, that human imagination, engineering skill and the ability to make a buck will combine to deliver new marvels that will make life ever easier. Humans have done it so many times before that it seems trite to suggest otherwise. Surely it is certain that we will put our fertile minds to the creation of new technologies, to many new things that we will use and, more significantly, sell.

Excitement takes over as we drive to the shopping mall to purchase the latest electronic game console. We may even get a little misty eyed, in awe of what the future may bring for interactive games that teach us strategies to make a virtual kill with cunning. But to our surprise we might also feel a niggle. It's just a slight twinge. You know, the sort of feeling that is buried deep down, well below our peripheral worlds. Only the feeling is definitely there and it's not all that good.

If we were to stop for a moment and search out this discomfort, we might find it translates into a question. Is there a cost? Yes, that is it. Is there a cost? Is it, this technological miracle of ever better, ever bigger going to last? Is it sustainable?

Sustainable. Now there is a word guaranteed to irritate. How can we be certain that the technology path that we have walked so effortlessly for as long as we can remember, will lead to greater achievement; to keep going, always upward? This is an important question at many levels.

We will defer to a later section the damage to our psyche of lives so connected to electronics that we never touch the earth. We will also keep away from the alternative world of organic spelt bread and hemp t-shirts. Not many really want the utopia of the technology bashers who can contemplate skinned knuckles from hand washing diapers. Here we will consider the practical meaning of sustainable.[4] Will the march along the technology path – the road to ever more innovation to

---

[4] The Australian Oxford Dictionary suggests '*able to be maintained at a particular level without causing damage to the environment or depletion of the resource*' as a definition for sustainable.

make our lives easier, more comfortable and entertaining – be maintained without damage to the environment or depletion of the resource?

This is the 'niggle'. We are unlikely to articulate it as a need for sustainability but what we feel is the fear that all the wonders we enjoy will stop. That we will clatter down the hierarchy of needs into poverty and destitution. We feel it as the fear of personal loss of job, house, security rather than sustainability, but it is actually the same thing. If we knew the system was sustainable then all would be well and the 'niggle' would go away.

There has been much spoken, written and claimed about sustainability with advocates, critics and agnostics all having their say. We will not go over that ground again, other than to note three things.

The first is that arguments for a sustainable future are contingent. They are all conditional or dependent on an event or events that are uncertain. For example, ubiquitous clean energy is possible, indeed probable, but we must have the technology breakthrough for it to become a reality. Similarly, recycling, reuse and frugality are things that people could adopt but only after they have made a strong enough personal commitment that their habits will change.

The second idea about sustainability is that demands on resources and the environment increase as a result of a growth in both the human population **and** in affluence. At the end of each day there are more people on earth than there were at dawn. In any number of ways most of the people already here have spent the day trying to better themselves. As a consequence of our desire to make more the demand curve continues to rise.

The third notion is that, as a general rule, demand and externalities increase together.[5] Whilst we can create technologies that will supply demand for goods and services

---

[5] In economics, an externality is an impact on any party not directly involved in an economic decision. In terms of sustainability, economic activity can result in damage or disruption to the environment that is not considered in the commercial decision. In effect the environment is not on the balance sheet.

with minimal environment impacts - clean energy from fusion power is an example - this feature is not a requirement of current technology. For those forms of production with a known impact, accounting for the cost of mitigation or rehabilitation runs counter to the business imperative to profit. Until recently we have decided to leave these costs off the books. The emissions trading schemes for carbon is a way to account for one externality, but there are many more.

Forms of production and technologies that use up resources or leave unpleasant residues tug on strands of the environmental web; ultimately these forms are not sustainable.

Let us consider what it will it take for sustainability to happen in the city of Sydney, Australia, that is home to 4 million people spread out in over a million dwellings across an area of more than 2,000 km^2. Every day resources are needed to feed, clothe, and provide shelter for these people to a level of comfort that they have come to expect as their right.

As a minimum we need 10 million kilocalories of food per day[6], 8 million litres of water just for drinking[7] and have used construction materials for the one million or so dwellings. We also need clothes that come in various shapes and hues to follow societal convention. Given that the people are spread out beyond distances that they can easily walk, most residents travel at least 10km per day using some form of transport with its own infrastructure and fuel needs.

Rather than send everyone out to forage for food and water, in our system a small number of people manage and secure natures harvest for the many. Food is grown far from where people live, then processed, packaged and trucked to stores that are convenient.

Most residents also have a strong desire to shop for items that make them look and feel good, so these products are manufactured too.

---

[6] Human calorie needs vary with sex, age and activity level and although there are several ways to calculate precise amounts for a given individual, a useful rule of thumb is 2,500 calories per day. However, many westerners would loose weight on this level of intake.

[7] Humans need around 3 litres of water per day of which 2 litres is taken as fluid. The actual requirement depends on sex, age, body weight, weather and amount of physical activity.

Just to think of sustainability in these terms should be cause for pause. That a modern city is maintained at all is a testament to the organizing power of the human mind and the systems that such thinking can put in place. All around the world cities exist and they appear to work. Every day goods and services are delivered to residents as demand dictates. Most of these cities are growing and we are led to believe that many are prospering, the engine of commerce driving a multitude of supplies.

Rather than try to do the sums for everything we will look at three: water, electricity and waste.

Domestic water use

Water use by the one million households in Sydney has become a contentious issue, especially since the last decade has thrown up more drought years than people expected. Currently domestic water supply for Sydney comes from a series of dams in the upper Nepean catchment, three quarters of it from one source, the Warrigamba Dam. At the end of 2007 these supplies were much depleted following a long period of drought, recovering somewhat with rainfall in early 2008. Even if the dams were full, demand is increasing at rates that will eventually outstrip supply[8].

The standard Sydney suburb where most residents live is made up of detached houses that sit on plots that average around 600 m^2 (0.06 hectares). Each member of the average family of four that would reside in such a house uses a significant volume of water for themselves. Each family member will take a daily 5-minute shower except for the teenage son who takes longer than seems necessary, and so consume 200 litres per day as they wash. At least another 50 litres are used for washing dishes and cooking plus 60 litres for

[8] Current water storage capacity is 2,600 GL and was last at this capacity in July 1998. As at 18 December 2008 available capacity was 1,660 GL with weekly supply to Sydney at 9 GL (or 468 GL per year). In the absence of recharge the city has about 42 months of supply.

the daily average of the laundry. Already we are up to 310 litres per household per day. The toilets, there are usually more than one, are flushed with potable water and in the older designs send up to 20 litres down the gurgler at a time - with three flushes a day, except for the son who may forget, the family will use 240 litres to sloosh away waste. Some simple addition and we are looking at 550 litres of water per day that must be piped into the house.

Extrapolate to all suburbs for a week and we have 3,850 million litres (3.85 GL), just under half the water supplied to the city.

What if the household decided that they would rather harvest their own water so that they would not have to rely on the water utility? This should be a more sustainable solution to their water needs. The roof area on a typical Sydney house is 250 m^2, so in the average year with 1200 mm of rainfall, 300,000 litres of water will run off each impermeable roof.[9] So if the water from the roof was harvested, stored in a tank behind the garage (or even under the garage if aesthetics were a bother) and then pumped to a header tank in the roof, the rain that falls in an average year is enough to keep the family clean, tidy and waste free for 78 weeks - a year and a half.

"Ah, but what if?" says the pub skeptic. You know, the guy who always finds a snag whatever the logic of the argument. "What if it doesn't rain on demand?"

This is true. Capacity to store 300 m^3 of water - roughly three times the amount in a good size private swimming pool - in a clean and safe way is unrealistic for the average household. Hopefully there will be some coincidence in supply from rainfall to replenish the water used, reducing the need for storage capacity. But we cannot be sure. Rainfall onto one suburban Sydney house made 20 mm only 5 times throughout 2006, and one of those was a 106 mm event that just about flooded everywhere including part of the basement. The inevitable feast and famine cycle that is the weather means that on-site capture carries a risk of shortage, a significant risk in dry years. There will be times when households will need to tap into a reliable

---

[9] In dry years rainfall in Sydney may be only 600mm halving the harvest to 150,000 litres, conversely wet years might see more than 2,000 mm of rain and a bumper harvest of 0.5 ML.

source. And this is one conundrum of sustainability; there will be times when big-ticket infrastructure is needed to maintain supply.

"Ah ha," the skeptic says with a grin "so how do you pay for the big ticket mains water supply if everyone is now using less water from the mains? Answer me that one."

Well we could raise the unit price for mains water.

"Ah that will just force more people offline, only the storage tank companies would make money."

There is a pause.

"And then what if you live in Wagga?"

Our skeptic has an endless supply of what ifs. This one refers Wagga Wagga, yes that is the name, a town in country New South Wales[10]. The average annual rainfall in Wagga (the locals usually drop a Wagga) is 569mm (22.4 inches), roughly half that in Sydney, and rainfall events are more erratic. Should residents look to onsite capture and storage? The sums are the same. The result is fewer weeks where onsite capture can meet demand. The good citizens of Wagga may need to rely more often on mains water.

"Which is totally unfair. Why should those poor buggers have to pay more? They don't have that kind of money" is the inevitable retort from Mr Skeptic. "Storage tanks, pipes, pumps, gutter filters, down pipe refits all cost money you know." He is right. Where infrastructure is concerned it pays to tap into an existing system with many users who, between them, share out the cost.

Who should, could or will pay to retrofit their suburban castle with onsite water capture? In Australia many residents, or more strictly the bank, own the title and so are responsible for any expenses associated with it, from insurance to insect pest control. So the owners would be expected to pay on top of their mortgage repayments for a water storage and pumping system.

Interest repayment on household debt in Australia averages 14% of income. We have borrowed up to the hilt. A spare $10,000 plus to be partially sustainable over water is a

---

[10] The appellation Wagga Wagga is thought to mean 'place of many crows' in the local Wiradjuri aboriginal dialect.

significant impost. It makes sense for the individual and for society to devolve this fundamental need for water to the collective. Not in the socialist sense as the system delivers from commercial payment for service, but in a collective nonetheless. It is easier to pay as you go.

In Australia there is money around for the water issue, including a $2 billion national water initiative plus input from state governments. But the cost of a $10,000 subsidy for homeowners and landlords to have onsite water harvesting is significant, roughly $10 billion for Sydney alone.[11]

"No way mate." Our skeptic chirps. "Why not reduce demand? That has to be cheaper. You know, shorter showers under a low flow showerhead, maybe every other day, half-flush button only, full loads in the washing machine, and wear your shirts twice. You could even put a bucket in the shower to catch the cold water that flows forever before the hot water arrives. Cheaper than a monster storage tank."

It would help to offset for a while. It might even be possible to reduce use by 20% with sufficient awareness, education and will. The problem is that frugality only provides respite. The problem will soon reappear because 50,000 new people per year move into Sydney. In a decade and a half we are back where we started.

"What about recycling?"

Now that is an idea my skeptical friend. Use grey water on site for non-potable purposes; shower water to flush the toilet; pipe the rinse cycle out onto the veggie patch or into a tank to wash the car. Sounds like a plan. Not too costly, minimal refits needed and immediate benefit.

"Sorry you can't." The skeptic says. "The regs."

Ah yes, the regulations. Grey water cannot be used in these imaginative ways because of any number of health risks. Bacteria can be very dangerous and this will be what the residents believe. Should one of those recycling attempts cultivate a nasty bacterium, an epidemic of gastroenteritis would cost huge sums in hospital resources, sick days and

---

[11] $10 billion is the size of the recent economic stimulus package extended to Australian lower income families as a $1,000 payout prior to Christmas 2008 - one response of the Australian Federal government to the global financial crisis.

public confidence. No this is too risky. We will have to keep the regulations.

There is something more significant. Should a scheme to encourage residents to harvest rainwater for more than just watering the garden be successful, then we could see the collapse of the major infrastructure. Revenues for the water utility would decline to the point where the service could not be maintained. This is especially true where the infrastructure is aging. Loose the power of the collective and the big ticket services are very hard to maintain.

At the household level this would increase risk. A reliable supply of water would be compromised. People would be very uncomfortable without a tap to turn on. It would be seen as a step backwards to times when people did it tough; a place where we came from and one we might remember with nostalgia, but not somewhere we want to go.

Domestic electricity

A few years ago almost all the suburban homes in Sydney would light up at night with globes based on technology no different to designs first seen in the 1800's. Incandescent light bulbs have given us light for decades but they are inefficient, delivering around 10 lumens of light for every watt of energy used. Over the years design tweaks have kept the cost of production low and improved reliability but failed to increase energy efficiency.

It took the notion that the risks of global warming could be mitigated if we reduced carbon emissions to herald a fast change. Almost overnight households changed to compact fluorescent bulbs (CFLs) that operate at around 65 lumens per watt, a 75% energy reduction.

The amazing thing was that it did not cost the householder anything to make the change. Companies bought the light bulbs as carbon offset credits and sent bubbly young people around with free CFLs that they happily installed. These cheery individuals also carted away the old, but mostly still functional, incandescent energy behemoths. And if your suburb was not on the rounds, show bags of free CFLs were available in shopping malls. Most houses are now illuminated with globes that are energy efficient and last up to ten times longer than the old type.

My 15th birthday was in 1976. I can't recall that there was much memorable about it other than I still had hair and had just saved up enough to buy my first album, the title of which I should not divulge. What I do recall is that it doesn't seem all that long ago. What also happened in 1976 was that the households of Australia consumed 2,371 petajoules of energy.

Next year my youngest son turns 15. He is already onto his third iPod, way past his maiden download from iTunes and, for the moment at least, quite happy to tell you it was *Dirty Harry* by *Gorillaz*. As he powers up his technology he will use a tiny amount of the roughly 6,000 petajoules of domestic energy to be used by Australians in 2009 – comfortably double the 1976 amount. When he gets to my age, another 30 years will have passed. Projections have household energy consumption in 2050 at 17,000 petajoules, another doubling and seven times the 1976 level.

Nothing in the sums that create these projections suggest this doubling rate will slow. Even though we can expect restraint, not least because over time energy is expected to be more expensive, energy use will go up because there will be many more households, enough to absorb an extra 12 million Australians. So to stay at 2008 rates of energy use, each household in 2050 will need to use 30% less energy than we do now just to accommodate the increase in households.

Can changing the light bulbs reduce domestic energy consumption by this amount? In houses lit by incandescent bulbs, 9% of the total energy use is needed for lighting. So a change to CFL will save just under 7% of energy use. Not enough. Households would need to make similar reductions elsewhere. People might swop plasma screens for the new wafer thin stick-on screens that some clever engineer will invent (TV accounts for 6%) or wear jumpers in winter and cop the sweat in summer (space heating and cooling accounts for 8%). Together these changes would save another 10%, but would still be 15% shy of the target. The big savings will come from having fewer showers (water heating accounts for 27%) and finding more efficient ways to keep the beer cold (refrigeration accounts for 21%). Solar water heating alone could make up the shortfall.

Will people purchase these more efficient appliances or modify their behaviour, live in very different types of house, shower less? Probably not, given that the evidence is that people consume more and more energy as they relax into affluence. Sustainability for energy is not something that can be worked out from the household up. The obvious solution will be to change everything, to use even more energy but from cheap, clean sources.

Domestic waste

A publication from the most reputable and helpful public service, the Australian Bureau of Statistics, suggests that in 2006, production of solid waste in Australia was equivalent to 1,629 kg per person. This is roughly the mass of a small car. The report also noted that half of this amount is recycled, whilst the rest goes to landfill.

Now if 800 kg of solid waste is crushed it probably occupies half a cubic meter given that a fair bit of the waste will be building materials and other heavy stuff. On this assumption and given that there are 21 million Australians, enough solid waste is generated each year to fill 3,500 Olympic sized swimming pools or, if you prefer, Sydney Harbour 21 times.

An Australian family of four generates around 77 kg of domestic waste each week[12]. Around 36 kg of this is green waste: peels, cores, rinds, skins, leftovers and green material. Not wanted in the garden, this host of readily compostable materials ends up in landfill. So much greenery is buried that companies are mining of methane from old landfill sites.[13]

It is much better for the family to compost at home. A 60 litre bin, suitably perforated yet closed to contain the odor, only

---

[12] This is 38% more than the average UK family.

[13] Whilst recovery rates may be profitable, the carbon cost of transport and fill to put the material there in the first place make the methane a poor carbon offset.

needs to have water and worms added for natural recycling. Layer in some brown leaves and the roses will be there to smell. It is amazing how much volume of decomposable material returns through entropy to tiny molecules and at a very rapid rate. The 30 degree days that are so bad for the evapo-transpiration do wonders for the decomposer bacteria. All they need is for it to be moist.

"Ah mate, mate," the big S says. "You want me to keep that stuff in my house. It will smell like Paddy's market in February."

A Tupperware bin with a lid, tucked away on the corner of the bench top will solve that one my man. Then the compost can go into the soil, even the smallest yard has some, or the space for pots that can grow some veggies. In the 1950's many households were self-sufficient in vegetables. Prepare the soil, plant seeds or seedlings from the nursery, add water recycled from the shower and the salad greens come thick and fast.

"Mate, no time for waterin', weedin', pesticides, diggin', more waterin'. Mate no time, we'll miss the soapie, then Idol's on and I'm away to work in the mornin' before it gets light."

Our skeptic has a point. Our lives are full and electronic distractions proliferate. More importantly, why would you spend time and effort growing vegetables when you can buy them in the store for a tiny fraction of your weekly wage? In the 1950's food was the major item of the household budget whereas today we can feed four people very well for less than 20% of the average wage – most of us spend more servicing our debt than we do buying food.

At first glance it would seem better to keep the decomposable materials out of the waste bins and recycle them at home because it would lower volume in landfill, reduce transportation externalities and ease the need to find suitable holes in the ground. The benefits to households are there if people have green thumbs, but generally they don't. We seem not to have the time, inclination or skills to grow our own food and there is little incentive when the shops are full of inexpensive, high quality produce. There are also risks to public health from pests who will benefit from the inevitable negligence that will befall some of the compost bins. The second glance promises little.

We also have the same challenge for waste as we do with water and electricity. By 2050 the extra 12 million people in Australia will generate another 2,000 swimming pools worth of waste to go with the 3,500 we manage to fill today – 12 more Sydney Harbors' worth. Whatever way you look at it we have a sustainability problem with waste that will require real innovation to ensure that reuse and recycling become attractive. The likelihood is that these new ways will require collective solutions – economies of scale work with waste.

Before we leave this wicked conundrum there is one further aspect of sustainability that is highlighted by green waste. The reason compost is so good for the garden is that it is full of nutrients that the process of decomposition makes available to plant roots. Perhaps a quarter of the green waste produced each week by a family comes from the kitchen; the peels, cores and leftovers. These nutrients were once in the soil on the farm that grew the food crop. Some of the nutrient originated in that soil, the rest would have been added to the soil by the farmer as various forms of fertilizer. What happens is a net transfer of nutrients off the farm. Some go into us as we assimilate the nutritional benefits of eating. Some pass through us and end up in the sewage system. And then there is the proportion that ends up as household waste – the nutrients in up to 10 kg of greenery per household per week.

Our ultimate sustainability challenge is to keep soils fertile. The best solution for green waste is to compost at a commercial scale and return the nutrients back to the soil on the farms.

Dodging the grid

'Restraint, reuse and recycling' has become a mantra for being green. We are green if we take shorter showers, put our shopping in reusable calico bags at the supermarket and compost our potato peel. We can be very green if we elect to purchase green power, buy a hybrid vehicle and offset our long haul flight to Bali. Some of us could even grow our own vegetables for our vegan diet, cycle to the office and never buy a mobile phone. All of these things are green and every little action will help us take some personal responsibility for our resource use.

What these things will not do on their own is significantly reduce demand for natural resources. All the human population growth models have the global population stabilizing at a minimum of 9 billion, a third more people than we have today. We are like all other organisms. Our desire to make more of ourselves is pervasive. It may not be possible to hold up population growth but we might be able to persuade everyone to be very green and reduce our resource use by 30%. This will not solve the problem but might accommodate the extra people on our current level of resource use.

Then there is economics. Even if it were possible for everyone to adopt green actions immediately, there would be a crash in demand sufficient to destroy the markets that supply the food that most of us cannot grow. Just like utilities, our food production systems need the collective to function efficiently.

The challenge is that our current levels of use are not sustainable. This is why green groups have been so vocal for the past decade. And the reality is that changing the light bulbs is not enough. We must change the light bulbs, change the power sources, return the nutrients back to soil, reuse and recycle our water and more. Whilst they require individual action, these tasks are not achievable as individuals. We will need collective systems with the right economies of scale that combine with an enhanced sense of personal responsibility.

Self-sufficiency for 9 billion people is unrealistic. What we have to invent are ways to make our water, energy and waste systems more efficient and aligned toward sustainability. We cannot dodge the grid.

Chapter 3

Story

The last loaf of bread

If I may, I'd like to tell you a story about Joe, a forthright and well-meaning man who was sound of heart and mind. Joe had a problem. The problem was this. He desperately wanted to conserve a loaf of bread. It is hard to understand why he wanted this, but he did. He wanted to save a loaf with a desire that made his stomach ache.

Now we are talking really preserve, keep, protect, look after, and perhaps even hug this everyday product. Clearly bread is for eating not preserving but Joe just couldn't help it. Something deep inside told him he had to save a loaf of bread.

It began suddenly one fine spring morning. Having bought a delicious fresh loaf of bread from a fine baker on the village high street, Joe had a feeling that this might just be his last loaf. Not just his, but *the* last loaf. It was a surprise but it gripped him like an attack of cramp. Maybe, just maybe, this loaf was the last of its kind.

Joe was overcome with a profound sense of duty to protect this oblong cob, this last loaf. Like some great turning of the tide it was inevitable. He had to do it. And yet he felt a quandary. The desire to protect was powerful, dutiful; noble even, and yet hunger was still the reason he bought the loaf.

Part of Joe still wanted to eat a slice or two of the loaf with peanut butter and jam. This was, after all, a very fine loaf; a loaf designed to be eaten with peanut butter. And here it was, in his hand, at it's very best - soft enough to squeeze between the fingers and yet with firmness that only the crispest of crusts could support. It even had the warm smell that first danced across his nostrils when he entered the bakery and was about to infuse his clothes and the deepest recesses of his Nissan Pulsar.

Perhaps if I were frugal, Joe thought, *I could eat just one slice a day and the loaf would last a couple of weeks.* This seemed to Joe like a reasonable plan. It would work fine for Monday and Tuesday, a bit hard on Wednesday, probably pretty stale by Thursday and altogether too risky to his health by next week. No problem. What, after all, were freezers made for? A freezer was the perfect repository to preserve a loaf of bread. So, our man, being smart, decided that he would pre-slice his loaf and put it into his freezer so that his tactic of frugality could be implemented without risk to his digestive system. Sure enough, the loaf would last for a couple of weeks.

Defrosting a pre-sliced frozen loaf is not quite the same as cutting a doorstep fresh from the cob. You miss the pleasure of rasping the blade across a crisp crust that still holds with affection the amazing soft centre, all set off with an aroma that just makes the mouth water and the lips say "more please". To pre-slice and freeze was a compromise, yet this loss of quality was acceptable to Joe. For all he had to do was retrieve a slice from the freezer, thaw for a few minutes at room temperature, apply the spreads and eat.

This was it; a logical solution. Yet it left Joe with a niggle, a feeling that maybe this was not satisfactory. After a week or so the loaf is finished. No more loaf. If this really was the last loaf ever, then surely there was something better he could do than to eat it. There should be some record of it as the last ever. Something to remember it by, a gravestone perhaps, or something more direct like the depressions in the cushion of grandmas favorite armchair. Perhaps the paper bag wrapper that the kept the loaf clean in transit and moist for its natural life could be framed and hung in a local museum.

Joe fumbled for options. Then in a surprising moment of clarity, one came to him. What if, he imagined, I preserved the loaf indefinitely? A big grin lit up his face. What an idea. "I am a walking genius!" he said to nobody within earshot.

The frozen option would do pretty well. So long as the freezer worked and the power stayed on, the loaf would indeed keep indefinitely. Sure, another food source would have to be found, and luckily there were some handy rice crackers in the cupboard. You know, the ones that line the calorie counting shelves of the supermarket and taste like cardboard - at least what Joe imagined cardboard would taste like. But this was a small price to pay for preservation permanence.

Joe made a decision. He must get serious about this cryogenic method and if he did, if he took a few precautions, he could indeed keep his loaf for what he thought would be forever.

So he bought a second freezer as backup in case the first one broke down and installed an alarm system to tell him if this happened. At the hardware store he bought an uninterrupted power supply system and, in case the power went off, the most expensive portable generator they had. Two jerry cans of petrol for the generator were stored safely in his garage. He toyed for days with the idea of buying a gas freezer just in case. In the end he just had to get one to stop the worry.

Fearful of risk and natural disaster, he moved his new preservation setup into a shed that he had built in his yard. Made out of brick and clad with the latest fire retardant, it was a very solid structure that included the latest in surveillance cameras, a digital alarm, an anti-tamper system and all the power backups for the electronics. Standing at the top of his yard, staring with admiration at his museum, Joe figured he had the bases covered.

A mate of his who was a risk assessor for an insurance company agreed. So barring acts of god or terrorism, his loaf was secure. Joe had made it safe and our man was pleased with his work. He could not, of course, eat the loaf, having gone to so much trouble to keep it. But the frozen loaf was preserved.

Many weeks passed and Joe began to forget about his loaf. He only checked on the shed occasionally because he was sure that the electronics were watching. Sometimes he thought about the loaf in an idle moment at the office and was glad he had installed such reliable surveillance; but by the time he got home the loaf was forgotten. He did show the loaf to his young nephew when he came to visit, although the little kid, who was still a year off kindergarten, didn't seem all that impressed and kept asking why he couldn't eat a slice with peanut butter. Joe put this down to the innocence of youth.

Time passed and Joe began to think that his loaf was important. It was now 15 years old, which is aged for a loaf of bread, and on the outside at least it looked pretty much as it did when first frozen. Joe's choice of modern frost-free freezer was a smart play; as was the idea of not opening the fridge unless absolutely necessary.

He began to imagine how valuable this bread would be to his nephew when the youngster was old enough to appreciate an aged loaf. Indeed, it would surely be worth a lot of money. He had read in the paper that a young lad was given a bag of coins by an ancient aunt that included a very rare dragoon worth thousands. Joe imagined that he could leave the loaf to his nephew in his will, but if he did, then there would be things to do about the really long-term storage of his loaf. What should he do to ensure its preservation until his nephew was old enough to appreciate it? It might not be easy to find someone with enough passion to step in and take responsibility for a frozen loaf of bread should anything happen, like being on the wrong side of a moving bus. He had got the loaf this far and he wanted it in the family, but it had to be safe. He didn't want to risk the impetuosity of youth.

He called his lawyer.

"In the future," Joe began, "in the future when wheat blight has made flour scarce, as it surely will, bread will become really rare. My loaf will be a national treasure. It would be the only intact, albeit frozen, loaf of bread left in the world. The only remaining example of a food that half of all humans used to consume every day." Joe paused. "There will be some value there surely. Maybe even a way to cash in."

It was in the nature of the profession not to question a client's motives and the lawyer calmly offered the solution. He would gladly establish a trust that would take legal ownership of the loaf and be responsible for its preservation in perpetuity. Joe accepted eagerly. The firm would prepare the papers.

Although he had the legalities covered thoughts of future rarity played a lament in Joe's head. He still felt uneasy about his frozen loaf. Something about keeping it in its cryogenic state indefinitely. *Why would that be good?* He thought. *There is no nutritional benefit and folk would only get to look at it very occasionally, perhaps only on television or on the inside page of the newspaper.*

In the nature of things that happen after a couple of glasses of wine on a balmy summer evening, Joe asked himself some important questions. *How did my loaf of bread come into existence? Why is it as it is? What am I doing keeping it past its sell by date?* And many other tantalizing teasers that were probably just the wine talking.

In the morning, still feeling a little uneasy, Joe sat at his computer and sent enquiries to the ether about the lost art of bread making. A lengthy search uncovered details from some old folk who remembered what to do. Flour, water, yeast and salt are mixed together in the right proportions to make dough, then some mechanical energy is applied with healthy kneading and the dough set aside to rest while the yeast does its thing. Bake the risen blob in an oven for a while and, almost by some miracle, a crust appears around a soft, aromatic centre.

Joe also found that you could grind a wide range of grains into flour that would produce bread with different texture and smell. It was even possible to add other ingredients for flavour, colour and nutrition and *voila* you had a whole range of breads. You could also mould the dough into a limitless array of different shapes before the miraculous intervention of heat to create baps, loaves, sticks, cobs, rolls, bagels...anything.

The most important thing that Joe discovered was that if you have the ingredients and an oven, plus some ability to follow the basic method, then you can make bread all the time. Joe imagined that it would not take that much to follow this method and to go back to his initial plan: to have enough bread to both keep and eat. If there was more bread being baked, Joe realized, then he could eat a couple of slices a day with peanut butter and still conserve his loaf. Not the loaf, but at least a loaf. Every few days there would be a fresh one to slice. It was very satisfying, not least because the loaf would be doing what it was made to do and not just stay in the freezer using up space and energy.

Just as Joe was about to abandon cryogenics he felt a twinge. Again it was that nagging feeling in the pit of the stomach, something remarkably like worry. *What happens if the ingredients run out?* Well, there might be a day without bread. Joe reassured himself that this would be fine so long as the baker can bake tomorrow. A loaf should be edible for a least a couple of days.

But what if it takes many days to get the ingredients? There could be no loaf baked for a week, longer. This situation felt risky and clearly violated the constitution of the trust. Conservation of the loaf with Plan A required that the process of bread making was secure. It was essential that the baker have an uninterrupted supply of flour, water and yeast, as well as power. The oven needed to fire every day.

Joe knew that he must learn more about the ingredients needed to make bread including where they come from, what happened to get them into the right shape for baking and how they got to the baker.

Several days later Joe had only managed to figure out the local supply line for flour. It was long and involved many more people than just the farmer who, thanks to a rural heritage grant, had begun to grow wheat engineered to resist the blight. There were a dozen or more folk involved in turning the kernels into flour and getting the right mixture into the oven. With more people, skills and crafts involved, it was obviously more complex to secure supply. It was one thing to know what the supply chain was, quite another to be in sufficient control of it to guarantee all ingredients and skills indefinitely.

Joe had several restless nights. He was now quite concerned about the future of his loaf. His return to Plan A looked shaky when he was awake and became a nightmare in the dark.

A small fortune was needed to ensure supply from the power corporation, the wheat farmer, miller, microbiologist and salt distiller, not to mention a cut for the many companies who transport these goods to the baker. A hefty lottery win would not be enough to buy all the land, equipment and expertise that were needed. The process was complicated, far too complicated. The freezer option might be the engineering solution, but it was much simpler.

Joe agonised for the rest of his life. A few years passed and he gave up on the resources to secure his own supply chain for bread making and resorted to the frozen option to preserve his loaf. In this at least, he succeeded.

The loaf had been frozen for 34 years when Joe died. It survived another decade under the trustees care until a share market crash sent the trust fund bankrupt, unable even to pay for the electricity to run the fridges.

When the Trust was wound up, no one believed that the loaf of bread had been frozen for longer than anyone could remember. Sure enough, wheat blight had eradicated flour from the human diet several decades earlier. The bakers sons were forced to other grains and finally figured out how to make something edible from maize meal. Bread made from wheat was ancient history.

Only the old folk remembered what bread tasted like. One wizened gentleman, apparently old enough, was asked if he recognised Joe's frozen loaf as bread, but he said "it wasn't twisty enough." Bread was always twisted in his grandfather's farmhouse. No one remembered what an old fashioned loaf of bread was like and, worse, this memory loss didn't seem to bother anyone. Like the dodo, no one had ever seen one or experienced its value. So it was decided that Joe's loaf was a fake.

There were portable DNA fingerprinting machines but, somehow, nobody bothered to check for wheat genes in the loaf. It was as though the passing of time had completely devalued this, the last loaf. Only the historians had any interest left.

Joe's nephew had become a civil engineer and enjoyed fortune and considerable fame as the founder of a company that revolutionized the design of desalination plants. He occasionally recalled the ingredients for bread, something that his uncle used to tell him when he was a kid, but he couldn't remember why.

Chapter 4

Story

Weeds

"It's a weed I'm telling you. Simple, friggin' easy, it's a weed."

"So what you mean is that if it grows naturally then it is a weed, right?"

"Well if it gets in the way it's a weed, natural or not. Anyway what's friggin' natural?"

"If it's supposed to grow there then it is natural and if it is natural how can it be a weed?"

"When the bloody thing gets in the way."

"Alright, have it your way Shan, it is a weed. Still doesn't get us around the problem that we can't clear it. Just because we call it a weed."

A young man with a ponytail turned towards the two farmers huddled around their beers. "Gents" he said, "couldn't help overhearing your conversation. I'm very interested in what you have to say. Can I shout you a round?"

"Who's interested?" Shan, the burlier one, grunted through a beard that had rarely been trimmed.

"Brad Thornley is my name."

"You're not from these parts mate," said Shan.

"No, I live in London mostly."

"Mostly?" said George, a slim individual but obviously of country stock who was still a little upset with himself for letting Shan win again.

"Well, when I am not here, which I am more and more these days."

"Alright, we're due a refill," Shan said, apparently not requiring the acquiescence of his comrade. "We drink VB."

"Coming up." The visitor moved with purpose to the bar.

'What are you thinking? Accepting beer from a tart with a hairpiece." Said George who lost mainly because he was always cautious.

"It's free isn't it? Usually you're so tight that we go home thirsty. Why waste an opportunity?"

"No mate, I make my shouts."

"Except when your missus cuts your allowance." Shan said as he slapped his friend on the shoulder, a little more firmly than the standard love tap. "Why would a bloke from London with girls hair want to know about our weeds?"

"Well, he must have some problem with it otherwise--"

"He wouldn't have asked. Maybe he's a rich bastard buying up grazing land to get mineral rights. Anything's friggin' possible these days."

"There you go Gents."

"Cheers Brad, Shan's the name." An iron grip was offered from the bearded one. Brad shook it firmly without hesitation.

"Cheers Brad, I'm George."

"It is a pleasure to meet you blokes. You have properties locally?"

"We do." George said. "Are you on the land Brad?"

"In a way I am."

"Cryptic mate." Shan said following his rambunctious instinct. "You on it or not?"

"Well, I don't own a farm, but I am involved with the land."

"So you're agency - although those buggers never buy beer for the likes of us."

"No, not government-" Brad laughed, "No, I work for an investment bank."

"A bank. Suits are a long way from the land mate. What you doing out here?"

"I have often wondered what investment bankers do." George said with a turn of his head.

"Well, mostly we get off on making money. Well, making money grow to be precise."

"So you plant twenties in the ground and harvest fifties." Shan smiled at his own suggestion. It was the nearest he ever came to laughing.

"It would be hard to do that. No, we do it the easy way."

"You nick it."

"In a way yes. We get people to value something more highly now than they did yesterday. Sometimes it means we have to improve it so it actually does obtain a higher value, but we really get a buzz from making something more valuable without doing anything."

"So how do you get the cash?" George asked, ever practical.

"Easy, we sell the thing when it is worth more than we paid for it."

"Bugger that. I'll chuck in farming and start a friggin' bank." Shan said. "I'll even get the dog to have a go."

"Sure Shan, there are some technical skills that your dog might find tricky, but the biggest thing is knowing how to handle risk. That's the hard part and the bit that stops most folk."

"This dog is pretty smart Brad, he figured out how open Shan's fridge and help himself, at no small risk to his safety. Has he got what it takes?"

Brad smiled. "Maybe he does," he said.

"Fleecing folk of their cash doesn't explain why you want to hear about weeds, natural or otherwise." Shan glared at George as he finished.

"Let's just say that I am curious about them," Brad said.

"Cryptic again mate." Shan sat back and folded his arms; dark hairs accentuated the muscles that made it possible to work the land.

"No. I am genuinely interested in why it's a weed."

"Well for starters it's not an 'it'." George said in a voice he saved for his budget figures that he knew in every detail down to the last cent. "It is at least four different species that sometimes get going on their own, or two or more at a time. Just depends."

"Depends on what exactly?"

"Oh, you know: where you are in the paddock, what stock you had running, how much rain, that sort of thing."

"Bollocks. You are crapping on again George. You get the friggin' weeds everywhere, whatever you do. Show me one of your paddocks where it's not everywhere weedy."

"No, some of that is natural regrowth."

"Crap. They're friggin' weeds mate."

"What do you mean by a weed Shan?" Brad said.

"Those bloody bushes that shade out all the grass. Blink and they are there. All you can do is rip 'em out."

"And they are native plants?"

"Not just native, them bastards are natural according to Professor George Wilson here."

"What happens if you don't rip them?"

"They grow a bit, then just sit there. Bastards just sit there and laugh at your stock trying to find some non-existent pick. Rip 'em I say."

"Shan's right Brad." George glanced across at Shan as he agreed. "They grow into shrub size, lots of them, and then tend to sit there. They don't seem to grow much after that, even when it rains."

"Can you burn them?"

"Not easily Brad, they don't burn really and they keep down the grasses that do. No, like Shan says, you can only rip 'em."

"Another round gents? Don't worry, its other people's money."

"Sure, thanks mate."

"Cheers."

A wry smile greeted the barmaid as Brad made his order.

At the table George returned to his quizzical mood. "So, what about this Bradley guy?"

"George you are an ignorant tosser. He's no more a banker than I am a male model. He's a greenie, just tryin' some back door tactics."

"You sure." George said, pathologically unable to agree. "He seems pretty kosher to me. I don't know too many greenies who would talk that way about making money."

"Exactly. George you are as dumb as the young 'uns. He's undercover; the bastards infiltratin'."

"So what do we do?"

"He's buyin' beer aint he?"

"I guess... I don't see we could tell him anything important anyway. We can't even agree what a weed is."

"Naw mate, you don't agree with me; me and every other bloke in the district know; a friggin' weed is a friggin' weed."

"Here you go gents. Isn't Danni a cutie."

"You should be careful coming on to barmaids out here Brad. Most likely they'll be related to Shan and he can be quite protective."

Brad smiled. "Thanks George, I'll remember that," he said then finished passing out the glasses and put the tray against the chair leg. "So has anyone left the weeds to see what happens?"

"Only if the bloke dies and he had no balls to make kids. Leave the weeds and you need a property the size of the friggin' state to be viable. Keep the bloody things and they spread to make you go broke in a hurry."

"Fair enough, but you blokes know your land. What do you think would happen if you kept them for say five, or ten years?"

"That's an interesting question Brad" George said. "I think that the weeds might stay for a while, like the cypress does, not growing much. Then maybe they'd get away and a canopy would appear."

"Has anyone tried thinning, like they do with the cypress?"

"Why would you? Bugger of a lot of work and still you wouldn't get any friggin' grass. Naw, chain 'em. That's the only way."

"Shan's right Brad. Most blokes need to bring on the pick and thinning is just too fiddly a solution."

"But maybe this stuff would grow back to some sort of woodland, maybe even forest."

"In who's lifetime mate?"

Danni came over to collect the tray. She smiled at Brad and allowed her hips to sway in an extra-wide path around the bar stools. "Thanks Danni" Brad said and returned her smile as she collected the empty glasses.

Shan glowered. "Like I said, in who's lifetime? Danni's kids?"

"Maybe it would take too long, but I've seen a bit of farm forestry around; what kind of rotations do they use?"

"Sure, I have a small hardwood lot Brad," George said. We rotate about a 16 year cycle but I reckon I need to push it out, recruitment to harvestable stock is slowing down."

"You wouldn't believe this bloke, he numbers all his friggin' trees." There was a hint of jealousy in Shan's voice. "He probably even has names for them all. Cyril, Cedric and Clyde the cypress trio."

"You know only too well that I grow redgums and blackbutt."

Shan took another long drink. There was a slight curl in his lip. "Mate, it's good of you to shout an' all, but what the hell do you want?" He said, ignoring George.

"Shan that's harsh mate, sorry Brad."

"Ah, no worries, it is a fair question." Brad paused. "I am just interested. Really. I drive around these parts a lot and have been, on and off, for a year or two and it's the same all over. The bush just looks like it's thickening up, and I just find that curious. Then I read a bit in the paper about what blokes were calling woody weeds, then when I overheard you blokes I figured I'd get some info from the horses mouth, so to speak."

"So you really are a banker then?" George asked.

"Yep, I am." Brad reached into the pocket of his coat and pulled out a gold case. "I don't usually hand these out in pubs, but here you go."

"Still don't tell us what you want." The beer was beginning to make Shan irritable.

"Says here Brad is something in charge of environmental markets." George said fingering the card to feel the embossed letters.

"So you want to make us into a friggin' market?"

"No Shan, I don't think I could, but your weeds interest me."

"How the hell do you propose to make a market out of them? Bastards just leave nothing for the sheep."

"That might be it, just might be it."

"Bollocks! You come in an' nick our land and we loose our livelihood; easy way to make money eh."

"That's not quite how it works but I admit the bank would want to make a quid."

"Buggered if the bank will make any more off our land."

"What about if they made a quid and you made two?"

"Don't take the piss mate. My old man took out a loan to buy our old place. When he finally paid it off he'd returned the amount three times over. He worked his arse off to meet the payments and that made him grumpy as all hell, but not half as grumpy as he was when I worked out how much he'd actually paid for the friggin' place."

"But now he owns the asset."

"Owned. He died a year after the final mortgage payment and left it all to me mum. She promptly sold it, went around the world with the local vicar and spent all the money on black kids. Now she's shacked up in the vicarage."

"Shan, you can't blame the bank for what Dora did with the fruits of your old mans labour." George said, which helped Brad as he almost felt obliged to say it himself, although he was composing a less personal reply.

"Bollocks. That's not the point. It's the fact that he had to work like a dog for thirty years to pay for the farm three times over."

"So what if you were to avoid that kind of graft Shan?"

"Sure, love too. Got the winning ticket?"

"Well maybe I have."

"Still cryptic mate. Gonna have to do better than that."

"I'm curious too Brad." George said. "I've heard blokes talk a bit about alternative markets. Is that what you mean?"

"Well you could call them that, but really it's just another way of gaining a return from the natural resource capital of your land. In effect you trade something that the land can produce other than crops or livestock. You get return for a lot less effort."

"Don't tell me we sit in the pub while the property just goes about making cash for us to piss away. If my old man was here he would probably take you out."

"Old man Doherty was like that Brad. He probably would have clouted you one. But if this market exists, then why isn't everyone in it?"

"Well it's like all new things. They take a little while to get going and to be taken on by enough punters for them to be viable."

"Ah, so whatever it is that you're crapping on about ain't viable." Shan latched on with glee.

"So if the market isn't viable Brad, why do they pay you to work on it?"

"Because it is viable, just not yet for small players. The corporate investors are making a killing because of the leverage they have to buy big. I'm interested in what can be done on smaller scales. Like on your properties." Brad paused expecting the obligatory interruption from Shan, but he was quiet twisting the whiskers at the end of his beard into a tight spiral. "Look, I've seen some scientists locally and maybe they'll come up with the missing ingredient."

"Missing ingredient? What would they know?" Shan said.

"I hear you. We don't usually pay much attention to them either, but I had some specific questions about those weeds. I wanted to know what would happen if... well, how to break that holding position where the plants seem to get stuck. The white coats were helpful but it needed some more research they said, as they always do."

"Well 'coz it keeps them in a cushy job obviously." Shan said jumping back into the conversation.

"There is that."

"So we have to wait while you figure out how to get us 'small players' into the game, the friggin' boffins to solve our biology and the big players to lay off from clearing out; then after all that all will be well and I'll be pushing up the daisys. Sorry if I don't jump up and down singing a happy song Brad."

"It's the nature of risk Shan."

"Bollocks. You can crap on about alternatives for as long as you like but the land is shot. It's too dry and it's been dredged of any riches it had. God knows how long the blacks burnt it, then we went and spent two hundred years taking sheep off it, trucking them to the cities and crapping the nutrients into the ocean. If that wasn't enough, the wind gets up and blows the topsoil away to New Zealand. Big surprise when all we get is friggin' weeds."

"There is truth in that Brad." George said. "He looks dumb and hairy but he's not – dumb that is."

"No, I could see that you blokes had a pretty good idea of what was happening and I agree that there are challenges, well, you know, difficulties. But the idea of markets is to reverse some of that loss."

"So what is the market selling?" George asked.

"The market trades in carbon." Brad said.

"Oh yeah, that's great. Customers in the deli we have a super special today customers, today only, carbon just $1.99 a kilo. That's right customers just 1.99, don't pass this one up, this price can't last."

"Well it is a bit like that Shan," Brad said unperturbed, "only the supermarket is a global one. Countries with space to capture carbon are selling that volume to countries who burn too much fossil fuel."

"Kyoto protocol Shan, we disagreed over that the other week."

"That's friggin' bollocks too."

"I can see what you mean, but the market might just be the way to recover the land. Make it productive again."

"You mean like a fallow Brad?" George said as he paid attention.

"If you like. That's a good analogy." Brad said although he was slowly running out of enthusiasm as more of his idea surfaced. "You get money for putting away some carbon as trees and in the soil, then the carbon makes the soil more productive. It's like being paid for taking medicine."

"Brad. There you are, I've been looking all over for you." A stunning blond in snug jeans and knee high leather boots glided towards the table.

"Gents, this is Gabrielle; my fiancé. Gabbi this is Shan and George."

"Hi guys, how are you goin'? Sorry but I'm going to have to drag this boy away."

"No worries, we've already fleeced him of all his beer money." Shan said with a grin.

"It's been good to talk to you blokes. Give me a call and we can do it again. Mobiles on the card."

Shan nodded with rather less vigor.

"It was a pleasure Brad" George said offering his hand. He winked and added "I knew they weren't weeds."

Brad took Gabbi's arm and they walked out of the bar.

"I leave you for five minutes and you are adopting half the red-necks in the district."

"Harsh my love, harsh. They looked it I know, and they were a bit gruff, but I reckon they know a thing or two."

"Like how to drink beer that no doubt you bought."

"Well they did know how to do that but, hey, redistribution."

"You and your redistribution."

"One thing though, you remember that Italian dude who kept quoting his business management theory – the financial mind, the manager and salesman couldn't all be the same bloke."

"Yeah, Safarelli or something."

"Right, him. Well those two are dead ringers. George wouldn't let an ant get past him and Shan would be able to tell your mother what to cook. All they need is a mate who could sell dope to a copper. I reckon they would make kick-ass business partners."

"Really. So how does that sit with your rule about not going into business with your mates?"

"It doesn't."

"So get your head out of work. You know my mother will freak if you talk shop with Dad."

"Me, talk shop! How could you cast such aspersions? As far as your mother is concerned I am the model son-in-law to be. She thinks the sun shines out of all my-"

"That will do..." Gabbi shook her head, smiled and drove the car into a driveway of white cobbles that made crisp notes under the tyres. Before the car stopped a woman in a pink apron was dashing across the stones.

"Ah Gabbi darling, so lovely to see you." The woman had her arms around Gabbi before she got out of the car.

"Mum."

"Hi Mrs B, something for you." Brad emerged from the passenger side carrying a huge bouquet.

"Brad, they're beautiful, you shouldn't have. What a darling boy. Gabbi, where did you find him?"

"You know mum." Gabbi said to her mothers back.

"Come in, come in. Your father is with the Doherty's out by the pool."

"You didn't say there would be guests." Gabbi said, louder this time. "I am not-"

"Oh, sorry dear. You know your father, stickler for formal numbers at dinner. On you go, I'll be through." She peeled off into the kitchen and waved at the direction of the hall.

Gabbi and Brad walked along a corridor through to the ornate formal lounge, where a baby grand sat gleaming with a fresh film of beeswax.

"That thing always cracks me up." Brad whispered as he aimed to run his finger over the woodwork.

"Don't." Gabbi grabbed his arm. "I told you, you can't touch it."

"Aw, you spoil all my fun."

"Gabbi! How is my girl, you look as beautiful as ever."

"Hi dad."

"Brad my boy, good to see you."

"And you Mr B. How's business?"

"Good my boy, good. Well, excellent in fact, we were up three points for the week and more than a dozen for the quarter - but you'll get me into trouble. Please come and meet my friends." He walked across to an elegant outdoor setting under a shade sail. "This is Kathy and Seamus, my daughter Gabbi and her fiancé Brad."

"Brad Thornley, pleasure to meet you Kathy; Seamus."

"Aye, pleased too, pleased too." Seamus said through a graying beard that did its best to hide a well padded face. The handshake was what might be expected from a frame as bulky as it was tall. "Bill has been telling us about you both, proud as punch he is."

"Oh, don't believe it all Seamus. He's right about his daughter though, wonderful isn't she?"

"Aye, that she is; that she is." He took a swig of beer. A dab of white froth attached to his whiskers. He wiped it the back of his hand. "Bill says you're into markets."

"Well that's what keeps me off the streets. Always been keen on new ways to make a dollar. Legal ways." Brad smiled. "Yourself?"

"Chemist, aye, soil an' all. Liked to find out what was in things see. Soil now, that's always got things in it."

"Amazing, I'm sort of into soil myself." Brad looked across at his future father-in-law to accept the wink he knew would be there. "What are you working on now?"

"Ah, some trace elements in fly ash added to agricultural soils."

"That's the stuff that comes out of power stations."

"Aye, that's it; thousands of tons a day there is. Gets buried in the ground. Always thought that was a waste. It's rich in trace elements see." There was another wipe of the beard. "Especially manganese and selenium. They can really get the crops goin'."

"Would it work on trees?"

"Aye lad, it would, it would."

"Seamus, Molly mentioned your surname is Doherty."

"Aye, it is."

"If you don't mind me asking, do you have a son?"

"No, but I have two nephews, Brendan and Shannon."

"Is one of them a local grazier?"

"Aye, that'll be Shan."

Chapter 5

Story

Johnno

"G'day."

"G'day mate, what can I get you?"

"Ah, a scooner of, what's this? Ah, yeah, Coopers, thanks."

"No worries." The barman smiled at his customer as he held the glass at an exact angle to the beer tap. "That's two sixty thanks." A gnarled set of fingers fiddled with, then passed over enough loose change. "Cheers." The coins found their way into the till. "Hey, you alright mate?"

"Sure, fine."

"Not seen you in these parts before, visiting?"

"From Burke."

"Ah, cotton country. I have a cousin from Burke. Went to visit him once, nice place."

"Yeah, nice place," the fingers shook as they reached around to clasp the glass. The stranger took a long draught of the beer. His face lightened. "Ah, better" he said "it's been quite a day."

"Bad or good?"

"Now that is a question. Not 'n easy one."

In the way of all good barmen, Jerry knew more than angles. "It's a quiet arvo, if you want to give it a go."

"Ah mate, it's pretty dreary stuff."

"Some of them make the best yarns."

"Well it's a yarn alright, but in the end I'm..." The stranger paused and his eyes glazed over. He blinked and wiped his brow, his fingers lingering on his temple.

Jerry watched and tried to guess the man's story. He was a shearer perhaps. Certainly the hands wore the scars of throwing sheep, but he wasn't bulky enough. His frame was wiry and beneath his musty looking sports coat, there was no evidence of the necessary muscle. So rural, but he was not a shearer. A drover maybe?

"I'm in town for a hearing on my case," the man said, "I had to appeal and, well, well it's kinda dragged on." Again, he stopped, his face contorted. As his hand went to his temple he tried to hide the pain; he blinked again and picked up his scooner. The draught of beer seemed to revive him as before. "Sorry mate, nearly lost it there," he said.

The young barman reached a hand across the bar. "Jerry's the name."

"Johnno."

The men shook hands. Although firm, the grip faded. Jerry held his end of the greeting long enough for the stranger to release first.

"High court?" Jerry asked. Many of his regulars came from the ornate building across the street.

"No mate, government. I'm challenging my PVP."

"PVP? What's one of them?"

"Sorry mate, Property Vegetation Plan." The stranger smiled. "Sounds pretty wanky, eh. And it is, really, although it's getting harder to laugh about it." Johnno took another gulp of Coopers, the once frothy head now draining down the inside of the glass. "You need one agreed by the Department if you want to change what you do on your farm."

"Sounds a bit controlling. I thought farmers choose how to go about farming." Jerry said.

"Kinda, I mean it should be. Only more like farce really. All I wanted to do was to have a working farm, a place that was productive without damaging the land. It seemed a reasonable request. My old man would always crap on about 'duty of care', that farmers were the custodians and doctors for the land. I used to think he was nuts when I was a kid."

"Was he?"

"Oh yeah, completely bonkers, but not about that. See my earliest memory of him was a massive pair of binoculars that he used to hang around his neck all the time. Dangling like some black badge. He would only let me see through them if I had the strap around my neck. First time I tried to hold them my fingers slipped and the strap almost chopped my head off. Scary thing for a five year old. See, my old man he was a mad birdo. It would take him days to plough a paddock stopping every ten yards to tick off another sighting. He wrote down the details of all the birds he saw every day for fifty years. It's quite a set of records actually."

"Sounds like a great hobby for a farmer."

"Yeah, he used to say that the birds told you about the land. Like they were some measure of how healthy it all was. I can still remember how excited he got when he saw something unusual. Supper time he'd come into the kitchen, eat the lamb stew, it was always lamb stew, and rattle on about this rare whistling chat or some wheatear that hadn't been seen for years."

"My old man would chatter too, only it was always about the footy."

"Na, we were never allowed to talk about that. It was birds or schoolwork. Feathered birds that is."

"Your dad still alive?"

"No, he passed on three years ago. He had a heart attack on his extractor. Still carrying those old binoculars."

"Sorry to hear that."

"Ah, he had a good innings, and he expected to go I reckon; especially since he was quite mad by then. See he had come become fixated with this bird, thing called a stone curlew. Funny looking bugger; bit like a scrawny chicken, but with eyes the size of dinner plates. The thing lives on open paddocks with stony soil, sparser the better. For about ten years my Dad couldn't talk about anything else, just this bloody bird. Obsessed he was. We had about half a dozen pairs on the property and Dad just adored them. He'd sit in his tractor and drive up to a nest, somehow the flighty things let him do it, and he'd just stare at them for hours. Don't know if you've stared at anything other than the TV for an hour Jerry, but it's pretty weird."

Jerry nodded and pointed to Johnno's glass, "refill?" he said.

"Sure, and one yourself."

"Cheers."

"So my dad, he gets so obsessed that he starts to manage the property just for these birds. He doubled the stocking rate on the best paddocks, just to keep the grass down. Then he stopped the fertilizer applications when he read that curlews don't like improved pasture, and I reckon he even fed them in the dry years. One time we had great rain for about a month, you know the really soaking stuff. The grass went crazy and my dad just panicked. He spent a fortune on fuel dragging a chain mower across each paddock over and over. Anyway all the frenzied activity must have done something. The number of birds went up to a dozen pairs. Dad would count the fledged chicks and was always boring the neighbours with questions about sightings on nearby properties."

"Sounds like he got pretty passionate about it."

"Possessed more like. It was freaky. I'd left home by the time it got really bad. My mom took the brunt of it. Bless her. She said he would get up before dawn in the breeding season and camp out at a nest the whole day just to keep the hawks away. Then when he did come home he was exhausted, would fall asleep on this chair. Never said anything to mom. He used to go down to the club once a week, but his mates got bored with curlew stories and avoided him. Sad really. Anyway he stopped going. At least on that night of the week he talked to mom, about curlews."

"Hard on her."

"Yeah, although she misses him I reckon. Never did understand that myself; gave up on women years ago. Couldn't find one like my old mum." The stranger smiled and a spark of life danced in his eyes.

"Hens teeth mate." Jerry happily agreed.

"Hens teeth." Johnno grasped the fresh glass of beer that had settled to an inviting gold color with a white top. "You married?" he said.

"Me, no, too young mate. Can't you tell?"

"Yeah, it was a dumb question."

"No worries."

"You're not?"

"Nope."

"Not that..."

"No worries. So your old man went a bit crazy, but how come you end up in a hearing?"

"There's the irony mate, right there I reckon." Johnno paused to take a sip. "See I was overseas, did a bit of work on oil rigs about the place and ended up a shift manager in the North Sea."

"I had you down first as a shearer or drover maybe, then as a farmer. Shows how much I know."

"Oh, done them too, when I was younger. It's hard not to when you come from the country. Maybe it was my old man, but I just had to get out. I went to Melbourne first, bummed around labourin' then met a bloke who worked for Texaco. Kinda went from there. Stayed away for nearly ten years, used to come visit in summer when the Pommies were here. Got married, then divorced and handed her most of my money. I was just getting back into it when my dad died. I figured mom would struggle to keep the property, so I came home to help. She really wants to stay home till she dies. Least I could do."

"Noble mate. Most blokes I know couldn't do that."

"Nah. I had just me, no kids, so it was easy enough."

"Still, good on ya."

"Anyway, I came back and the property was in a mess. Nothing worked, we had no animals of our own, but all the paddocks were hammered."

"How does that work?"

"Easement stock by the truckload. Basically my old man was feeding all the sheep in the district, just to keep his birds happy. Mom didn't know too much about the land, and she was pretty crook. I spent a week walking the paddocks and all I could see was dirt. It was weird; I didn't recognize the place even though I had ridden over every stone for twenty years. Worse part was the soil was disappearing down gullies big enough to swallow a truck. Buggered it was."

"Was? You managed to fix it?"

"Well, it's still buggered thanks to the hearing. See a mate had come across a neat way of recovering hammered land by rotating grazing, they call it timed grazing now. Apparently the idea came from Africa, but it worked a treat on Mike's place, so I figured it might save me Dads. So I spent the last of my dollars on some mobile fencing and brought in some cattle. Sheep don't cut it apparently, it has to be cows."

"Yeah?" Jerry raised an eyebrow.

"Ah, it's something to do with cutting up the turf. Not that there was much left. Still, figured, better follow the system."

Jerry held up his hand, then moved down the bar to serve two young women who kept talking to each other even while they ordered and a pink drink for a man in a business suit. Returning he said "Wouldn't that stuff up the birds?" One of the women overheard and glared at Jerry.

"Yeah, they hated it. So I kept a couple of paddocks for them. I didn't have enough fencing to do them all anyway."

"Sounds like you got it sorted."

"I thought so. Even mom started to feel better. On her good days, she'd come round with me to help move the cows."

"So what happened?"

"The bloody greenies got me. Somehow they found out what I was doing. I guess they knew easy enough as my dad was always getting visitors wanting to see the birds. Apparently Dads' paddocks are the only place in the state where you're sure to get to gawp at the bloody things. So when the birdos saw the fences and big mobs of cattle they were ropeable. One woman kept phoning me up and threatening me. 'You can't do this, you can't do that.' You know the type. It was like listening to my ex-wife. It went on for weeks, but I was buggered if I was going to stop, so one of them, probably the crazy woman, dobbed me in to the council. Not that I could see I was doing anything wrong, just trying to make a go of it again."

"How come the Council had any say?"

"Turned out they didn't, but they did tell the Department of Conservation who sent the heavies round. Some inspector. A right dickhead he was, said that I had to have a PVP done otherwise I'd get a huge fine."

The combination of alcohol and ire was potent. Johnno now sat upright and alert, shedding the slouch he wore when he first sat at the bar. "Same bloke comes back a week later with a computer and maps of our place. Have to say it was clever stuff. I mean, some of the tech we use on the rigs is pretty nifty, but this thing was up there. It was neat to see the satellite image and zoom in to each paddock. And up to date. You could see the lines from the cows across the first two paddocks I had only started rotating a month before."

"So what does one of these PVP things do?"

"Ah, its some planning tool. It's supposed to get the best conservation outcomes from farm management. The maps were clever right, but it would take a pretty smart map to know my old man's land better than he did. When you cut the crap though, the PVP just limits what you can legally do on your land. I guess it's handy for sorting out those stupid buggers who insist on clearing more paddocks, but most blokes know that's dumb."

"Sounds like you weren't into clearing anyway."

"Nah, turns out it was those bloody birds. See, some up themselves scientists on some committee in the city just decided that the stone curlew was now an endangered species in the state. A few months ago they voted to put it on some list, so now it and its habitat suddenly become super protected. You can't do anything that would threaten them or alter their habitat. So the whole of my dad's property would have to stay the same. I had to take away the fences and put sheep back."

"Even though it was your dad that made it look like that?"

"Exactly." Johnno leant back on his stool grateful for the recognition. "Pity those idiots in the hearing didn't get it. Mate, I'm telling you, I'll end up more nuts than my dad trying to convince them."

"So..." Jerry leaned closer, his eyes focused on Johnno.

"So, the PVP puts up a red light. Literally flashing red blobs on the screen and the smart arse gloats like he's scored. You know that smirk a customs officer has when he's going through your bags." Jerry nodded. "So I'm supposed to ditch the cows, get rid of thirty grand worth of fencing and bring back sheep. Which is why I'm here 'cause, I was buggered if I was going to cop that."

"The hearing?"

"Yeah, it's the only thing you can do once the fancy system flashes red."

"Appeal?"

Johnno nodded. "Only I have to spend me own time and good money to come down here to put my case. Ah, the rich stud owners and cotton barons send their lawyers who have bureaucrats for breakfast, but your average battler don't have that kinda money or connections."

"So the hearing didn't go well?"

"Well, I thought I had them. See I did a bit of digging before hand. It turns out this curlew is as common as muck just about everywhere a few hundred kilometers north. There are millions of the buggers. It's only endangered here 'cause it's too far south. I even called some Uni expert who wrote me a letter confirming what all the books say. I think it took the committee by surprise. There were two blokes and a battle ax of a woman to hear me. One of the blokes, a Dr Somebody, knew I had them but he said nothing. The woman got shitty and just kept quoting the regulations at me. 'These are the rules Mr Fogerty' she kept saying. I just kept on explaining why I had to save the farm and what had to be done to it."

"Good on you mate" Jerry interrupted, "I want to know what you told them, just let me serve these customers." He moved along the bar to the woman with the glare, hastily filled her order and delivered a pink refill to the suited man who had been trying, unsuccessfully it seemed, to chat her up. "Sorry mate," he said, leaning again on the bar opposite Johnno. "So tell me about this grazing thing."

"Just like I told them, it's about getting back good ground cover. Basically let the plants get going before they get grazed to buggery. I read this fella's book, the one who figured it all out in Africa, Savory is his name, and his book says it's about getting the roots going before the grass is grazed too much. You need to get the plant stores up before you let the animals eat, and then give the plants plenty of time to recover; lots of grazing free days. That's why you need fences. You move the cows around in one big mob and they stay in a paddock for only a day or two. Funny, but they soon get used to it and are real easy to move. They know when you open the gate there will be fresh pick, so they line up waiting for you."

"But the birds don't like it."

"Nah, they get their nests disturbed by the cows, but mostly they don't like the ground cover that grows back when the cows leave. Can't see their food I guess."

"So what did they say?"

"The committee?"

Jerry nodded again.

"Well, nothing in the end. Come back in a month they said, but meantime no more rotations. Which just about kills it. If I can't move the cows around, then the whole game is up."

"Look Johnno, this doesn't sound right to me. If you're up for it I might be able to help."

"How's that mate? Free beer."

"No, but I can get you some publicity."

"Hold on. I'm a simple bloke from the bush, not some peroxide cricketer. What would I do with publicity?"

"I'm talking about national newspaper broadsheet."

"How..."

"In my day job I write for the Fin review. Editorials."

Johnno leant back on his stool, his eyed widened, and a smirk curled its way across his face. "Ah, hidden talents my friend." He said. "Why the barkeep?"

"Oh, I do a couple of nights a week to get out of the flat, and it's just the best source of stories." Jerry smiled widely.

"So you'd write a piece?"

"Sure. Your story deserves a voice Johnno."

"Well, I told it to you easy enough. Probably no harm in it." Johnno reached to pick up his beer then stopped. "Hold on," he said. "No one wants their nose rubbed in it, especially the ministry pointy heads. That's not a good idea, it would kill the slim chance they will see reason by the next meeting. Don't mate."

"Didn't you say it's all over without rotation?"

"It is, but..."

"So you don't have much to lose. In a month it's all over anyway."

"Maybe."

"I'm good at general. I can keep your identity out of it. But, hey, no pressure."

"Mmmh" Johnno grimaced, "nah mate. Thanks, but no thanks."

"No worries. Think on it and give me a call if you change your mind." Jerry passed across a business card from his shirt pocket.

"Is that Jerry?"

"Speaking."

"Mate, it's Johnno."

"Sorry, I don't recognize your voice."

"John Fogarty, sat and spilled my guts to you in your bar a few weeks back. Said you'd like to do my story."

"Ah, yes, Johnno. How you going?"

"Bad mate, real bad. If it ain't too much, I'd like you to tell that story for me."

"Sure, sure." Jerry paused long enough for an uneasy silence. "Trouble with rare birds wasn't it."

"That's it mate, curlews."

"So, you want me to write something?"

"Mate if you could. It would be for my mom. She's crook... Cancer."

"Sorry to hear that."

"Quacks just found it. They gave her a year, maybe less..." There was a pause. "I just want her to live out her days at the farm. So I need to make the property viable again or if I create enough stink I can keep her there long enough. Ah bugger, it's desperate mate, desperate."

"Didn't you say there would be another hearing?"

"That's right, it was this arvo."

"So you're in town."

"Yeah, in your pub on the same barstool, only a good-looking lady served me a schooner." An attractive blonde in her early twenties looked up and smiled.

"Ah, so you have met Penny" Jerry said and raised his eyebrows at the phone. "Look, stay put. I'll be there in ten."

"Thanks mate, appreciated."

Jerry sat on the stool next to Johnno. They shook hands and Johnno insisted on his shout, which Penny served with another pleasant smile, but not quite Jerry's precision pour.

"So Johnno, what went on?"

"Mate, they nailed me. I kept the fences and the cows going after the last time, but the greenies got their binoculars on me, and dobbed me in again. Same surly bastard came round, only I refused to let them onto the property. A fine came in for that, two grand. Then, they sent an order to desist or risk criminal action. I was real mad, but I looked at my mum and I had to do it. I took the fences down and sent the cows back. I even invited the inspector back to see that I had complied but he didn't come. They just sent me a summons an' today I wasn't allowed to say anything at the hearing because I didn't comply with their first ruling. Bugger that, I didn't know they had any legal powers anyway."

"And your mom is sick."

"Yeah mate, real crook. Stays in bed or just sits watching TV, then, twice a week I drive her to the hospital." Johnno took out a handkerchief and wiped his brow. "So I'm stuffed. Only thing I can do is look for another place for mom and go back on the rigs to pay for it. I could put the farm up, but no-one with any sense would buy it. Anyway it's all tied up in Dad's estate."

"Well, sounds like some publicity might help. I'll write a story, no worries."

"Thanks Jerry."

Jerry looked at the man in front of him, his elbows perched on the bar seemed to be the only thing holding him on the stool, all energy had gone from his body and his eyes had retreated into the recesses of their sockets. His right hand was shaking.

"There must be some appeal procedure." Jerry said. "A bloke I know works for big law firm, give me a minute, and I'll see if he can help."

"Above and beyond mate, much appreciated."

"No worries."

Jerry punched his mobile, calling first his lawyer contact, then his publisher, wandering between the bar and the cigarette machine as he talked. After ten minutes he came back to the bar. "OK Johnno, so far so good. It's easy to appeal, and there's probably legal aid to help. More important, the record on appeal is good. Most folk seem to get over the line. If you can, it will mean sticking around for a couple of days."

"No worries. A neighbour's looking after mom and she won't mind if there is a chance of mom staying home."

"I'll talk to the manager, get you a deal on a room."

"Mate, thanks. Just one question."

"Sure."

"Why?"

"Why?"

"Yeah, I'm a stranger that you don't know from Adam. Why do this for me?"

"Well first I believe you, and your situation sucks. It's not what should happen. It's a classic case of the tail wagging the dog and that is dumb. Also I just want to help and..." Jerry stopped, his eyes looking up at the wooden rafters above the bar. "And I keep getting a weird vision. It happened first when you were telling me your story before, then a couple of times over the last few weeks, and then strongly tonight."

"Woah. Not some freaky hippie type are you? Coz you don't look it."

"No, I'm sane, least I think I am, but I keep going back in my mind to an experience I had in Africa."

"Africa? Coincidence that is, I did some time in Angola myself."

"Yep, it was nearly twenty years ago in Zimbabwe, the Zambezi Valley to be precise. I went to a research station, part of the student exchange program and had a blast. Elephant, lions all the usual safari stuff."

"Yeah, it's great."

"Just the one image I keep getting was from a day in Mana Pools National Park. The Wildlife guys were capturing the last handful of black rhinos and trucking them to private game reserves in South Africa. They said it was the only way to protect them from poachers. All the patrols they could mount hadn't stopped the killing. So, I'm in my late teens and I'm standing on a wooden boma made of mopane logs, and below me, in this cage less than ten-foot square, is a male rhino. This close he's the size of a bus and he's so pissed at waking up in prison that he's given himself a bloody nose trying to bash his way out." Jerry closed his eyes, then breathing deeply he went on. "This huge animal raised his head and looked right at me, and even though these things are supposed to be near blind, I never felt a more intense look in my life. It was hot with anger and intent, but mostly with injustice – a deep soul-searching sense of injustice at being in that wooden cage." Jerry almost felt the intensity again as he opened his eyes and looked across at his companion.

"Heavy stuff mate, it got to you. I remember. Africa does that."

"Absolutely it did. And I remember it often but never so heavy as when you told me your story. So I have to help you, otherwise I'll keep seeing the spirit of that magnificent animal, caged in the name of conservation." He stopped again, lifted his beer glass and peered into it before taking a drink. "If I was honest it's actually why I became a journo, figured that if I tell folk about injustice, it is the best way to deal with it."

"Then the spirits put us together mate. And I'll drink to that." Johnno raised his glass.

"You and me both... You and me both."

The two men talked for another hour, the beer steadily lightening their mood. Johnno recounted stories of wild binges that made the rigs bearable, testosterone fueled antics that Jerry matched with excesses of his own. Soon talk settled on extended recollections of Africa. They found they had shared delights, experiences and a sense of belonging to that place that neither could explain.

After more rounds than either of them could remember, Johnno staggered upstairs to a murky but comfortable room, bars being the original hotels in this part of the world. Jerry somehow managed to walk two blocks and climb the stairs to his flat.

A door opened.

"Johnno is that you?"

"Yes mom, it's me. How are you?" Before the old woman could answer he went on "gee you look great" and he reached out to embrace her.

"Don't be ridiculous, I'm crook and you know it."

"Mom, none of that talk now." Johnno said with a smile that faded as he turned to hang his coat. Dot caught the change of expression.

"Son, stoke the fire can you."

"Sure mom, just give me a minute to fix the kettle."

In a moment the fire was blazing again, flames beginning to lick the form from an ironbark log that Johnno placed on the coals.

"So," the old woman said.

"Mum, everything is fine."

"Don't try to kid me son, I can see it in your eyes. Old and crook I may be, but I still know you."

Johnno looked around from the fireplace to his mothers rag like body, slouched into the sofa, held up only by cushions in rainbow colours. A knitted shawl tucked in lovingly around her feet.

"This is new." Johnno leaned across and unnecessarily rearranged the shawl.

"Mavis made it for me. She is an angel that one."

"I'm glad."

"So am I, now tell me."

"Mum..." Johnno's voice cracked, only his eyes were able to communicate. He lowered his head onto the shawl.

Dot forced herself forward. She winced but reached out and placed a hand on her son. "I remember." She said. "When you were a boy...." The energy from the touch came to him. "Tell me Johnny" she spoke to the small boy at her feet, running from his father in awe, yet afraid of his wrath. Johnno lifted his head. He saw his mother young again, with endless peace and calm to pour over his troubles. "Tell me," she said.

"I, I'm sorry mom, we..." he finally managed to find the courage to meet her gaze. "We have lost the farm."

Chapter 6

Coda

The awkward news is that we are users

In this section we have considered the idea that humans are programmed, just like all the other animals and plants, to be users. We have an insatiable desire to make more. Consequently to use less is not as simple as it may seem. Conservation of the earth and its natural resources seems essential but is a huge challenge given our success in more-making. There are now very many of us, each with that built-in desire. And as our success continues so the human population grows along with a need for resources. We are users first and foremost.

We say that we are inventive, the best of all problem solvers. We will invent and organize ourselves through the challenge of our resource needs. Adaptability is our saving grace. This cheers us up, wards off depression, and gives us hope for our future.

Certainly our technological advances are staggering. There is no reason to doubt the capacity of the invention juggernaut that has put us on the moon, kept us alive though all kinds of disease and put the power of information access into our palms. In fact our technology seems likely to get bigger, faster and ever more pervasive.

What we have explored is that problem solving may be what we do, but we have been selective about the problems we have solved. We have helped ourselves to become ever more voracious users and we have not always foreseen the consequences, such as the inevitable increase in appropriation of primary production and use of land and water.

Joe showed us one consequence. It was that the things we take for granted might not be with us forever, even when we adopt extreme measures to hold on to them. Solving problems of resource use for 6.7 billion will require that we leave some things behind.

Even with sustainability, we will not always have everything that we have now. We can retain processes but it is very hard to retain objects. Joe also showed us that, in a changing world, it does not make sense to freeze things in time.

Whilst we invent rapidly our psyche is less supple. Our imaginary farmers, embroiled in an argument over weeds, did not have the full range of experience needed to engage with a world where natural resources have become decoupled from the growth of finance. No longer is money tied to a tangible asset[14], it has a life and growth of its own. Although the financial expertise found our farmers and explained what could be done, the leap of understanding was a big one. Not as big, perhaps as the leap of faith to take action and change what they had done for generations. They were cautious, and, for good reason, their reticence helped them manage risk. In the end though, there will be money to be made from multiple and alternative uses of the land.

We may be innate problem solvers but this does not translate to knowledge.[15] Our cleverness does not guarantee systems of production that get the most out of the environment, especially in the long term. We get the best out of a paddock for ourselves with little thought that there are consequences. Sometimes, when the creek silts up or we notice another species in decline, we seek a solution for the consequences. This solving

---

[14] In the past paper notes were normally freely convertible into pre-set, fixed quantities of gold (The Gold Standard). Currently all governments use fiat currency, money that has no intrinsic value (or guarantee that it can be converted into gold or another currency), other than by government order (fiat), it must be accepted as a means of payment.

[15] Eckhart Tolle makes the point that most of our leaders in business and politics are clever, sometimes very clever, but not intelligent. They are highly effective in using their brains to get to a result that has only a narrow frame of reference.

is a reactive action, and just as an aspirin relieves the symptoms of a headache, it is not a cure.

Johnno thought he was doing the right thing by his family and by most measures of common decency he was. He committed time, effort, financial resources and his own sense of justice to help his mother live out her final days peacefully in her own home. The system failed him. Johnno became a victim of the inevitable conflict of interest that happens when we try to implement sustainability. Johnno and his family were caught in the conflict between green and brown. In the real world there will be many like Johnno who will need help to adjust. Their stories will touch our souls, pique our collective conscience, and allow us to make the social changes that must accompany our invention. And so dissolve the conflict.

The real message and awkward news is that we must use resources. It is in our genes to make more. If we try to use less it will not be simple for us, perhaps impossible. However, as our resource appropriation increases, there will be a struggle with our conscience, so that some of us will accept having less, doing without and restrict our innate desire to want more. It may even be possible to implement some sustainability.

'Users' will also tell us that even if we use less, it will not be enough. We cannot meet future demand simply by saving. Our use of natural resources will have to go under the sustainability microscope. We will see that our current systems of production are unpleasant, inefficient and, in the long term, not sustainable at all. There must be a shift in the way we do things – and not just "green" or "brown" things. All things.

Part Two

Touch the earth

Chapter 7

Overview

Touch the earth

Disconnect

Nature can be quite disagreeable. There is little to commend wind chill sending the temperature below freezing; no fun in the attentions of angry bees; pain when caught outside in a hailstorm; real danger when faced with a bushfire.

Then there is also the inconvenience of it. It is hard to get up on dark winter mornings. Those pesky flies are a nuisance. So is the dirt that the dog brings in when the kids leave the back door open. And let us not get started on the weather. What a bother.

Then on the other hand there is comfort. As an old camping friend used to say when he pulled his luxury folding chair from the boot of the car 'No point in roughing it." We love our comfort and nature can upset that. It is one thing to sit outside in the shade on a summer day with a cool breeze blowing. But come dark we retreat indoors to a safe spot on the couch. It is better to be comfortable, a desire that lurks at the root of our urge to make more. It costs less energy.

Humans have become incredibly adept at screening out nature. We cannot imagine flies hanging around the Doctors surgery, or rain interrupting a trip to the photocopier, so we create spaces where the insects and the rain cannot go. In fact we are the masters of creating nature free space. It is a big part of our success.

We wake up in our soft, warm bed; then saunter without fear to have a shower. The water is always good and we have it at just the right temperature. We eat our breakfast from packets and bowls, put on clothes appropriate for the season and catch a glimpse of the weatherman on TV to see what the weekend looks like. We do this because this is the only time of the week when it might actually influence our plans. Walking through into the garage we start our vehicle and set the climate control. After parking the car, as close to work as possible, we may briefly encounter the day before entering our air-conditioned offices - which ironically we will set to 24°C in winter and 18°C in summer. If we're lucky we may even have a view outside. When we get home it is close to dark, we eat, watch a little television to confirm the latest weather and catch a police drama, and then return to our warm bed.

Given our lifestyles, it is no surprise that the bulk of us are blissfully unaware of the planet we inhabit.

A recent example of this occurred at the hairdressers, a place where we reduce our remaining natural protection from the elements. I asked the snipper; "What do you think about this whole 'environment' issue? Do you know much about it?"

"No, not really" was her reply, "I'm not really that interested."

Not interested? This being the same 'environment' that supplies you with air, water, food, clothes, buildings, that nail polish, your family, the family pet... and you're not that interested. I was aghast.

Then on a second thought I realized she had a point. All the things I had listed, along with a thousand others, have simply arrived in this person's life. They are a given for her. She has interests no doubt, but they are not to do with where all her stuff comes from. She knows that it comes from the store and she obtains it with the money she earns from cutting hair.

There are also people who are scared of nature. Not just the discomfort of prickly things; but a real fear. Sometimes it is the panic when we see animals like snakes and spiders or the dread of lightning when we hear thunder. In some, this fear is intense enough to stop them from enjoying time outdoors. Whilst this is a shame, it is wise to be somewhat cautious when surrounded by nature, particularly in parts of the world where things can eat you.

The reality is that the chances of coming to grief in "the wild" are very low. Despite robust populations of highly venomous spiders living in the towns and cities of Australia, deaths from spider bites are extremely rare. On average shark attacks result in less than 2 fatalities per year, the media frenzy after each incident notwithstanding, and another 2 people die from snakebites. Lightning strikes kill between 5 and 10 people per year. Yet in work-a-day Australia around 380 people are murdered, 1,600 are killed in road accidents and 2,000 die in a work related accident or work related disease every year. Statistically at least, it seems a lot safer in the bush.

The fear is real though and the statistics can always be questioned. No doubt we see fewer accidents in nature compared to town because so few of us spend any time there. We are too busy trying to cross the road to get to work. The interesting thing is that we accept daily work accidents, road fatalities, pollution, homicide, stress and toxic chemicals as cause for the early demise of our brethren; yet we are frightened of nature's perils. Heaven forbid that a shark might kill one of us in the water, once in a very blue moon.

If you deny your lungs oxygen you will die within three minutes and not make it to the end of this chapter. The oxygen that enters your lungs as you read comes from only one source – plants. It is true that rocks contain oxygen, and so does water and lots of chemicals, but virtually the only way you can get it into your lungs is via the miraculous structure of living plants that convert it into a gas. Have you ever noticed that there is only one 'e' between plant and planet? Perhaps that is because without plants we would not exist.

Water, food and shelter are also variously extracted, grown and sourced from nature and the environment. We know this. Indeed we have known since kindergarten that the environment is where we obtain the fundamentals we need for our survival. It seems strange then that we don't display a little more interest in the system that supports us. This is unfortunate because it suggests that we do not understand what is happening in the environment and our tendency is to fear the things we do not understand.

The reality is that we are a part of nature, even though we have learnt how to adapt it to our purpose. We are in it, like it or not. And yet the screen is up and so effective is it that we have become disconnected.

Connection

When people take a walk in the countryside, away from buildings and roads, their brains work better. Research at the University of Michigan explored the cognitive deficits caused by a short urban walk. Marc Berman argues that after spending a few minutes on a crowded city street, the brain is less able to hold things in memory, and suffers from reduced self-control. The research also found that spending an hour interacting with nature, like being in a park, helps improve memory and attention by 20%. In fact, just looking at pictures of natural scenes can improve brain function.

Despite our apparent indifference, most of us understand this concept of connection on an intuitive level. It just feels good to be outdoors. We have all felt the wind in our hair at the beach or on a hilltop view and felt strangely satisfied. As children we are fascinated by ants, pets and insects and rejoice in the feeling of sand on our legs and grass in our hair. This contact with the earth thrills us even as we are told by our parents not to get dirty and to put our grubby clothes in the washing machine. Then, quite suddenly we "grow out of it" and so for most people in the western world at least, we live out our lives decoupled from nature.

In the 1990's a not-for-profit environmental group, Conservation Volunteers Australia, developed and ran a program called Green Corps. The premise was to provide young adults with an opportunity to gain practical experience and training in the environment industry, and at the same time benefit many stakeholders with practical work outcomes. One of the Team Leaders recalls his experience with a group of 17 to 20 year olds.

We all got along pretty well, but there were quite a few in the group carrying issues related to broken homes, teenage pregnancy, abuse and drug use. At a point early in the program I had a training day and the organized activity had fallen through.

Being unsure what to do I took the team for a drive to a near-by nature reserve thinking that I'd be able to find something to talk about. I stopped the vehicle in a non-descript patch of bushland, sat on the ground and took out my sandwich to eat. Except for those needing a smoke, the team mostly stayed in the vehicle and chimed

"C'mon boss, can we put the radio on?"

"No."

"C'mon boss, can we go?"

"Not yet."

"What are we doing here?"

I was beginning to wonder the same when I noticed two ants were battling over a crumb from my sandwich. "Check this out," I said and within minutes the whole team was outside watching the drama unfold.

A commentary of sorts began and when a third species entered the fray there was uproar. One of the team discovered a spider eating a wasp and the afternoon was spent examining and discovering processes that occur in forests, especially those themed around conflict.

By the time the course ended six months later I could barely stop the vehicle before the doors were thrown open and the team set off to explore and enjoy the bush.

There is great power in "getting your hands dirty". Something magical happens when we place our hands into the earth. As the research on brain function suggests, we benefit when we identify with living things. It explains all the camping, gardening and outdoor sports.[16] And yet the screens that we build to keep the environment at bay are robust. In our daily lives we do our very best to ignore nature.

---

[16] In the UK there are more paid up members of the Royal Society for the Protection of Birds than there are paid up members of the main political party and they spend a staggering £70 million on wild bird food every year.

My wife works in an open plan office. She came home one day with the astounding news that indoor plants were not allowed in the room. "Bollocks to that," said we and smuggled in a birds nest fern, which quickly became a star attraction. Work colleagues now argue over who gets to keep it on their desk when my wife is away. And since the fern arrived another illegal immigrant has infiltrated the nature exclusion zone.

Concern

Opinion polls suggest that two in three Australians are concerned about the environment, a statistic that has been creeping upward for a while. So far the democratic process has not converted this awareness into concerted action. Funding for stewardship of the environment at all levels of government is tiny compared to welfare, health, education and defense.[17] When it comes to spending taxpayers' money, we spend it on ourselves.

What about the money in our pockets? About a fifth of the average Australian household budget is spent on food and non-alcoholic beverages, a fifth on housing, a fifth on transport and a little less than a fifth on recreation. The remainder goes to household furniture and equipment, clothing, various goods and services, medical expenses and 2% to domestic fuel and power. Beyond an amount spent on food there is no direct payment to support the environment that sustains us. As with the tax dollar, we spend our own money on ourselves.

Perhaps we do not see the need to pay for the environment. One explanation for the lack of concern from the hairdresser in response to the direct question about the

---

[17] In the Australian Government 2008 budget, social welfare was expected to cost A$97.8 billion, health 44.4 billion, education 18.4 billion and defense 17.7 billion. The public service was to cost the taxpayer 19.6 billion but expenditure on Agriculture, forests and fisheries was just 3.8 billion. A category within the agriculture budget for 'natural resource development' was set to spend 766 million, down 10% on the previous year.

environment is that many of us trust that "the government" and "experts" out there have got her best interests in hand. These important people are on the case so she doesn't have to worry.

Not according to most green lobbyists; those environmentalists who claim to speak for the environment. In the 200 odd years of European style occupation in Australia, the system of government has presided over resource use that has reduced most of the rivers to drains, put billions of tones of topsoil in the ocean, including some blown as far as New Zealand, and wiped out plant and animal species at a rate faster than anyone else. No, she should worry about the government, they are not the forthright stewards she, or they, think they are.

The trouble is this. Our aforementioned hairdresser's life has not changed because of any of the above. In fact, by most social and economic measures it has improved out of sight. Compared to her grandmother, who also cut hair for a living, the times are much easier, conditions more comfortable and there is very little to suggest that any of this will change. In fact, in material ways life just gets better. And even if there have been a few problems like global warming, it wasn't her fault.

It is clear that we touch the earth with our footprint, even if we interact little with nature. And for some of us our personal fears keep us from nature. This is a combination that makes it hard for us to see the truth. What we don't do is touch the earth with our perception.

Ignorance

We have been spectacularly successful in our quest to adapt the environment. Our choice has been to satisfy our needs and to shield ourselves from real and perceived harm. We do this because the primal urge to make more is alive in our subconscious. It is as true for those about to purchase a new air conditioner, as it is for those not sure where to find the next meal.

We choose ourselves in a simple sequence: me first, my family next, then other humans close by who look and act like me. After that, we choose others far away, but only in extremis. We do not think about what such choices mean for natural systems that supply the resources. Some might claim that they do think about it but the evidence of our actions shows that these thoughts are fleeting.

The consequence has been resource use on a grand scale and a litany of environmental woes with more to come.

The standard jibe of the green movement is to label these woes as 'deliberate negligence' or 'the fault of the government' or 'the greed of big business'. This is disingenuous. Any environmental woes exist because too few of us relate to nature. We do not feel what is happening to our planet and so we do not understand, let alone act on its behalf. We are blinded by our compulsion to make more.

It is fascinating that none of the Ten Commandments and precious little of any religious text, have much to say about the environment, how it works or why it may be sensible to care for it. Yes, you can find references to it, but in the body of text, environment is a small concern. And arguably, when the texts were written it was.

Perhaps if we understood more about what we risk loosing, then we would be more careful in our choices. But most of us don't know. Our understanding of the environment is poor. Even the ever-popular nature documentaries tell us more about the wonders than the worries, save perhaps for the impending loss of a rare and endangered species.

Alternatively we don't care because too many other more important things hook our attention. Either way, our modern society is missing an environmental ethic.

There are two important reasons why we might gain an environmental ethic to infiltrate all we do, and say, and are.

The first is the deep personal benefit to understanding nature. What it is, how it works and the truth that we are in and of it. This knowledge will not only make us feel better but will open the door to wisdom. Through understanding we can truly value what the environment gives us in our quest to make more.

The second is that an ethic would help us avoid the pain of our resource appropriation. There have been consequences of adapting the environment to our needs on an unprecedented scale. So far we have been shielded from these consequences yet we are rapidly approaching severe pain, collective and individual.

Scholars, soothsayers and mystics alike have always sounded warnings of environmental catastrophe. We don't like warnings, so we tend to ignore them. But these are interesting times. The fact that the internet search engines register one request a second for information on global warming suggests that more and more of us have a hunch that there is something important afoot.

As sentient beings we have the ability to learn and the ability to discern value. Given these great talents we can describe and agree to a set of moral principles on our use of the environment. We will always be users but we can be users with a conscience. It is very important for us to touch the earth.

Awareness

Modern humans live in a very different place to the one our ancestors inhabited. They spent their days in the open, walking in search of food, water and shelter and their nights gazing at the stars. Today light pollution dims the stars for half the global population, those people that live in cities. Few of us even see the cows that provide us with milk or the chicken that lays the eggs we scramble for breakfast. If we did see then we might change our production practices, but we don't. Chicken and bacon are flavours, not animals. We are disconnected. We do not see ourselves as users.

One simple action is to become aware of where our food, water and services actually come from. Not just the way nature provides them but also the organization needed to bring the produce to our tables.

It is easy to become aware of nature. It is all around us. Take a walk among the trees and start to feel what that is like. Begin to feel the 'isness' of all things. Run your fingers through the soil.

Wait. There is no reason to do this. We may not be connected but we also have no reason to be connected.

Everything that we need is given to us without the need to physically touch the earth. It is even possible to be aware without losing comfort. We are not going to make things harder just so we can connect. No, touching the earth is a personal choice.

The remainder of this section explains in essays and stories why we don't perceive and aims to show how hard it really is to make the choice to be responsible. We may touch the Earth. We may even feel the connection and be motivated to reduce our impact. But, then again…

Chapter 8

Essay

Taking your time

Nature and time

Gunung Batur is one of several volcanoes on the island of Bali in Indonesia. It has erupted more than 20 times since 1800, including in 1917 when a stream of boiling lava flowed down the mountain, only to stop just short of the temple at Batur village. The molten rock killed 1,000 local people. Earlier eruptions were even more dramatic covering large parts of the island in up to a metre of ash. Whilst devastating at the time, these deposits of nutrient rich ash soon combined with the tropical climate to create ideal growing conditions for plants. Human ingenuity then leveled the land and diverted the streams to flood fields of rice, the staple food of 3 million human souls who now live on the island.

The force of a large volcanic eruption devastates life in the nearby landscape. Nature is suffocated by hot ash or consumed by molten rock. The land is laid bare. What happens after this shock is why we marvel at the world in which we live. The land will be barren only for as long as it takes for the rock to cool, for almost immediately the wind will bring in microbes and fungal spores together with dust laden with nutritious nitrogen, phosphorus and trace elements. These essentials stick to the new surface and life starts. Really there is no such thing as barren if you are small enough.

The miracle is that on what looks to us like bare rock or inert ash, nature settles in to metabolize and make more. Nature gets going with the construction, development and growth of an extraordinary carbon factory complete with inputs, outputs, physical infrastructure, workers and management. A productive process fueled by energy from the sun and greased by rain.

First the microbes settle and, when the moisture levels allow, proliferate on the tiny packets of nutrient attached to the dust. They work away at the chemistry of the rock too, slowly accessing more nutrients locked up in the basalt and ash. Mobile creatures perch to give wing muscles a rest. Some, that are already dead, are blown in on storm clouds. Seeds arrive on the wind or in the waste of the birds and bats. In the geological blink that follows the lava flow has life.

As time passes, more dust accumulates, weathering assists the microbes with the breakdown of the ash and the biology proliferates in the cracks and depressions. Different microbes and fungi arrive and flourish on the first arrivals or their byproducts, and the conditions begin to change. In time it is possible for some of the seeds to germinate. This is a turning point because with the arrival of plants a new process begins that captures carbon from the atmosphere to convert it into the raw material for plant tissue. Most of these early seedlings will survive only briefly, as it may still be too dry, hot, windy or cold on the exposed surface. They will persist mostly on the energy reserves in the seed, but even as they die, so they add some resource to the factory cleaners. Whilst they survive they are green beacons that attract birds and provide some sustenance to insects that are blown the same way as the seeds. Soon there is enough metabolic activity going on to shift carbon from the air into the organisms and their by-products[18].

This metabolism repeats. Again and again, until more carbon is captured and, via the cleaners in the decomposition system, adds to the dust. This accumulation buffers the

---

[18] In our modern human world where carbon emissions are counted, we call this process carbon sequestration but we used to call it primary production. Secondary production is where animals consume the plant material to make more of themselves.

microclimate and before too long there is soil - a vital mixture of the inert and the alive.

On an island like Bali where it is warm and wet, this process of soil formation can be rapid. Cracks in the lava start to fill and even though the conditions may still look harsh to us, seeds of plants able to tolerate such extremes arrive and flourish. They grow, set seed themselves and die or spread via runners along the crevasses. Now when insects are blown in there is something for them to eat; so they proliferate too, creating in turn a resource for bigger animals that eat them. Sunlight, water and nutrients are all that nature needs to get on with what it does best; making more nature.

Plants make more through photosynthesis, steadily pulling carbon out of the atmosphere. Animals respire, processing energy fast and dropping wastes as they go. Some animals respire so fast that their lifetimes cannot be anything other than short and their spent bodies become nutrient packets to add to the waste. These byproducts of rapid more-making are also a source of carbon and nutrients used by detritivores[19].

Once it has begun, this chemical factory, or more strictly a vast 'industrial estate' of chemical companies, just keeps going - water and warmth permitting. Nature also fuels the chemistry of moist soil that works on the rock to release more nutrients making the cracks larger and deeper. Water and wind provide some disturbance, added to by the physical force exerted by plant roots.

Fast-forward several thousand generations of plants setting seed, animals feeding and bacteria working on the waste and there will have been a succession of chemical factories on the lava flow. This sequence of different types and activity of organisms covers and changes the lava and ash beyond recognition. New chemical operations will have arrived that are better suited to the changing conditions than the first arrivals: grasses, shrubs and trees that create both cover and structure. New animals move in that like this diversity of food and physical space. In the end something quite stable emerges that

---

[19] A set of bacteria, fungi and animals that are the recyclers that find homes in the moistening soil of the cracks, depressions and crevasses

covers the layer of ash. This overall process that results from more-making is predictable. It always happens and is ongoing, even though at times it can be interrupted, slowed or set back by new disturbances. This is a climax to the process of succession and an end point of sorts.

This process is a given. However, the exact makeup of the plants and animals along the way and at the end point - the specific species that occur and in what combinations - is not certain. Chance, the sequence of arrivals, the local pool of available species, the weather, and disturbance to the sequence all have an effect on the final outcome. Succession may be certain but not the form that it takes.

Succession is inevitable because, along with the vagaries of climate, as nature makes more so it alters the substrate on which it sits. For example, some of the carbon atoms that entered the plants through stomata in the leaves take a journey that sees them end up in soil humus. This is the important material that holds on to moisture and nutrients. Some components of the humus can be bound onto the lattice of clay minerals. This carbon becomes essentially inert again in the depths of the soil. Yet clay with carbon has different properties to clay without carbon, holding on to nutrients and moisture more effectively; a change that starts small, but is cumulative and eventually creates new extremes of wet, dry and nutrient availability for plant roots.

Once conditions change so does the balance of power between the organisms; some factories benefit while others are stretched even by small changes to moisture, temperature and nutrients. Eventually some are forced to close down and, in the gap, new opportunities emerge.

Scientists have struggled to understand the complexities of ever-present change, even though the basic principle is well known. Returning to the carbon finding its way into clay lattice, it would be easier to predict what will happen if this was the only effect of photosynthesis, a neat physical change of linear form. It would be nice if A+B=C, but because A *interacts* rather than adds to B this gives a chance for C to interact with A that brings D and F into the picture (E was on vacation), except that, by chance, G and H are there and open to an opportunity[20]. In nature the interconnections are many, chance inevitable, and synergies strong. We are left with the knowledge that change is certain and that succession follows change but on a pathway with many loops, forks and dead ends.

As time passes and succession proceeds there will be times without rain or times when the rain arrives as snow and fails to melt. There will be times when the temperature rises and rates of evaporation increase. At these climate driven interruptions the rate of more-making slows. Nature needs a drink.

Also if it is cold then nature slows down, a function of a physical law that says atoms move faster when it is warmer. Nature likes her comfort too.

So where there is rain all year around but strong seasonality, the rate of more-making is dependent on temperature – most scientists live in these places. Where rain is rare and/or ice abundant the process of more-making is driven by moisture – fewer scientists live in these places.

Even the rigorous observers of nature's more-making only see the tiniest part. The ecologists and naturalists who make a living from seeking to understand nature observe only that part that is in front of their nose for the blink of time that they are around to witness. Because even when the rain is patchy and the wind cold, more-making happens, just more slowly than our own more-making respiratory engines can perceive. More-making happens in the water deep underground, in the blackness of the deep ocean and under thick layers of polar ice. It even happens on our skin, home to a huge array of bacteria, fungi and mites, however often we shower.

What we do see is that the sliver of planet earth between the sea and the sky contains an uncountable number of metabolic engines run on carbon, water and energy from the sun. It is that simple. So simple in fact, that we figured it out a long time ago and teach the basic idea to preschoolers.

---

[20] Chance creates opportunity for organisms because wherever interactions create a new pool of carbon, so life comes along to use it. The reason that viruses that cause infection in humans become epidemics is because there is a resource (many humans) and the opportunity (humans are close together).

The amazing part is that the carbon engines not only stay serviceable but they adjust to workload, conditions and the supply of fuel. Through more-making, filtered by natural selection, nature copes elegantly with all the bumps, jolts, scarcities and bounty that come along. Tsunami, drought, earthquake, late frost, fire, flood or volcanic eruption; it matters little. Disruptions may be fatal for individual organisms but to the collective carbon engine, it is just part of the way the engine works. And as nature copes so it generates huge complexity.

Thanks to more-making and succession, so the lava flow and ash fields become colonized and transform. Barren for only a moment, they now have a blanket of soil supporting plants that are sucking in carbon dioxide to make more of themselves. And thanks to the great synergy of nature, these plants support the creatures that eat them, who, in turn, also make more. It is as inevitable as time and taxes. What was briefly barren becomes lush and diverse. All it takes is nature and time.

A thought experiment

Imagine yourself standing among some trees. It can be in the mature forest that we now know should cover the lava flow on the slopes of Gunung Batur or in a place that you remember well. Take yourself there for a moment.

It is early morning and there is a cool breeze beneath the shade of the canopy that leans across above you. Pungent wisps of tannins, turpentine and phenols emerge from the earth to tickle your nose. You cannot help but hear the twitters, squawks and buzz of winged creatures going about their business of respiration without a care. It is a nice place to be.

Take a deep breath and inhale some carbon. Feel that magic moment when your whole body pauses, just after you exhale slightly more carbon than went in.

Take some more conscious breaths and you will slow down.

Ease into the smooth tempo conducted by the joint batons of temperature and moisture and to a tune played by the organisms. Just telling yourself to breathe will make your own engine slow to something closer to the rhythm of the place[21].

It is a great truth to find that nature is relaxed. Stress is not in her vocabulary, she just slows down if conditions get tough; little wonder we seek natural places to help us relax.

Stay with that pure rhythm for a dozen more breaths and feel the gentle carbon exchange going on all around you. Hear the hum of the engine. Watch the branches sway gently in the breeze, held up by the solid sentinels of the tree trunks. Catch a glimpse of the birds foraging through leaves or scratching at the ground. Sift your nostrils for the scent and flavour of the flowers, bark, soil and pollen. In this imagined place you cannot catch hay fever.

Maintain the wonder of the observer. Keep your breathing deep and we will start the experiment.

We are going to press the fast forward button. You stay as you are, breathing gently, but every time the button is pressed time speeds up. Each press condenses real time into ever-shorter packets that relate to your gentle breathing. Here goes.

Button pressed once. Nature is scooting along and completes an hour of activity in the time it takes you to draw in one lungful of fresh air and gently let it go. As you take this breath, so the flies have buzzed past your nose a hundred times and are visible only as a hazy cloud. A pigeon has pecked at invisible seeds faster than a piston in an engine. A flower head has moved slightly to follow the sun. Save for a wave or two through the branches the trees have remained still.

Keep breathing gently. After another three or four breaths you will notice that the pigeon stops pecking and takes a quiet moment to rest on a low branch. Then he preens vigorously before flying again to find more seed. A bird party arrives in a cacophony of calls and movement on an intake, only to disappear as you exhale. Life for mobile creatures seems very hectic indeed.

A few more breaths and the flower starts to close because the light is changing before your eyes. The heat goes out of the day and the colours get warmer.

---

21 At rest most adults average around 12 breaths per minute.

Relax. Remember to breath slow, conscious breaths and we will push the button again. Now one breath sees one day of forest time slip past. One inhale is enough to see the daylight and one exhale the night. The birds and bees are no longer visible. They move too fast for the eye to see and the brain to compute. The sun skids across its arc splashing a burst of light and warmth through the canopy. The flowers wave at you as they follow its path. The trees seem to be breathing too, responding to the warmth they stretch and open their leaves to soak up carbon, then close them up again as you breath out.

Take a breather with the trees. It is too much to try and keep up with the animals. It is clear that they respire at far too fast a pace.

You may notice a little dampness for an instant during one of those breaths shared with the trees. Not to worry, that was just a shower and it is dry again now. Indeed at this pace it is dry much more often than it is wet because rainfall is really an occasional thing in most parts of the globe. Organisms get wet only briefly, although you notice that your feet stay damp much longer than the rest of you because the earth does not let go of the moisture so easily. It may take several breaths without rain before the dead leaves and mineral soil dry out. More likely it will rain again.

Now that the animals are just like specks on a poorly tuned television and we are in the natural rhythm of the trees, the button is pressed again.

You are still standing in the grove, breathing gently, but now a breath lasts one year of actual time. In ten breaths, a minute in your perception, a decade passes. The flies have been, completed their more-making and gone in one breath. Most of the smaller birds will have done the same in three or four contractions of your diaphragm. But at your feet the earth is moving, fidgeting with the input and decay of dead leaves raining from the trees. There is a wave of cold air followed by warmth pulsing around you like a heartbeat.

At this pace the trees move. Branches reach out like arms taking an item off a high shelf. Wrinkles and cracks in bark move as the trunk visibly expands and after twenty breaths you will have watched the trees dance as they climb ever higher above the pulsating dance floor of grasses and shrubs at their feet. If the trees in your grove are deciduous, at a point in every breath they will flash yellow, live naked for a moment, blossom, and then turn green.

Let go and feel this new rhythm. Just as you blink the place is black. The soil is laid bare as the briefest of bushfires consumes all in its path. Only the highest branches are spared, but before you realize what has happened the trees are green again thanks to shoots appearing from nowhere to clothe the trunk. But during your next breath these new branches fall away again. This was the trees response to a disturbance; grow lots of temporary branches in a hurry. Whilst the trees are shrugging off the effects of the fire, the grasses, herbs and shrubs have gone wild on the pulse of nutrients.

It is worth a longer pause to see nature at this pace, a little meditation in a world that we do not know. It is possible to watch the trees grow with the energetic flush of youth, then mature to produce seed in great profusion which for some species is during every breath and for others only occasionally in their mast years. You can observe that most of the time, most of the seeds go nowhere; at least no new saplings appear out of the ground. Then, without any obvious warning, recruitment happens and there are so many saplings that they look like a wriggling mass ready to swamp any other plants. But then, breathing a little more, you see that not all these youngsters make it past saplings and even those that do are not all destined for the canopy.

Then after our attention has been away just briefly, we look back and see that the parent tree is not as sprightly as it was. Some branches have given up and fallen to the ground, melting into the earth in a couple of breaths. In just a few minutes of your time the tree has sprouted, grown, reproduced and then falls in a storm. This accelerated view of nature tells us that the cycle of life – birth, growth, reproduction, death, decay – is the same whatever the organism. There are millions of variations in pace and timing but the sequence of more-making is always the same.

This is about as far as our thought experiment can go without stretching perception beyond our ken. We are, after all, around for a shorter time than most trees[22]. So perhaps we should stop now.

But then again, it might just be worthwhile to push the button again, just to see.

Still standing in the grove, relaxed, one breath now covers a decade. The life cycle of the trees are now easier to see. Saplings appear and thin out like ants foraging across the ground. Storms, fires and dry years occur in an instant. All the animals have long since blurred into insignificance and even the trees, solid and still when we started, now have a frantic look. They shoot up and spread skywards, quivering wildly only to pulse and wither.

A dozen breaths and we have passed a millennium - time enough for humans to have gone from riding horses to flying in airplanes - but our grove still looks the same. Sure, it pulses now that we can see the trees growing and the ground vibrate with the passage of detritus; but close your eyes for a moment and you will open them to roughly the same scene. Except that, wait, it is not quite the same. Yes, a few more breaths will confirm it. The young trees are not the same as the old ones. New species are arriving. Maybe the canopy is a little more open than it was and there is a little less dampness in the air. These changes are subtle, even at this accelerated pace, but they are there.

At this point we are starting to see and feel beyond the lifetimes of the organisms to long term processes that nature moderates, including the natural and somewhat predictable change in the composition of species that covered the lava flow we described earlier. Ecologists recognize numerous variants to the basic type but as one breath now allows a decade to pass, we can actually watch succession happen.

We can also see other processes. Soil creep, imperceptibly slow to our normal senses, exposes some rocks and covers others, steadily making any steep slopes shallower. If you imagined yourself in a tropical forest then this soil

---

[22] Whilst most westerners can now expect to live 75 years, such longevity for the majority is a very recent phenomenon. In 1900 life expectancy at birth was 50 years and prior to the industrial revolution around 35 years, even less where conditions were harsh. Reductions in infant mortality have contributed much to this change.

movement may still be too slow to see but if you imagined yourself in a young landscape like the one in Bali, the floor of your forest will have shifting foundations. We can also begin to see climate changes, the effects of weathering, and other mechanisms that shape the natural world at rates too slow for us to perceive as we go about our everyday lives.

If we are lucky though, we might witness a more sudden shift; a phase of fast change when the forest moves from one structure to another. This might be because of a steady increase in temperature or rainfall frequency, perhaps a reduction in the number of frost-free days or its opposite, an external disturbance or an intrinsic change such as a defoliation event. Whatever the trigger, the forest now grows with a different species mix and into a different structure. But just as you may hope to see a lion on a safari, there is no guarantee, and such a shift may not be seen. In fact it is more likely that the forest, pulsing with its yearly cycles and animated by longer timeframe processes, just ticks along, solid and secure, alive and ever-present, resilient to disturbances.

It would be easy to think that breathing tied to decades reaches the limit of ecological time and covers all the actions of organisms in their more-making as there are few organisms that eke out their livelihoods beyond a hundred years. However, this is not everything ecological because ecology also covers the changes in populations of organisms with a raft of theory as to how numbers might change over time.[23] So it will be necessary for us to press the button once again.

Breathe now and a millennium passes. A thousand years with a lungful of air. It is a struggle to imagine the processes that operate at this pace. All that we have seen up to now is going far too fast to be visible to our senses. The lives of individual organisms are lost in a fraction of a second. Only a handful of trees with exceptional longevity might be noticeable as entities. The rest of the scene will look a bit like a lava lamp of slow moving blobs passing around us. What we observe is the

---

[23] Formally termed population dynamics, this is the science of numbers of individuals within a species. Combine the population dynamics of all the species in a location and it is possible, at least in principle, to describe succession.

dynamics of plant populations - the cycles of boom and bust that are driven by external change and moderated by internal interactions. According to the classic theory, there is growth in numbers when resources are plentiful thanks to the obligation of organisms to convert resources into more-making, followed by a slow down in more-making as resources are depleted and become scarce. The consequences of competition for resources are fewer seeds (births), poorer survival of seedlings (young), earlier death from nutrient limitation (starvation), increased disease especially from overcrowding - all contributing to lower life expectancy - and a tendency to disperse. The instinct to make more remains and individuals still try, but when resources are scarce the attempts are less successful.

Theory is elegant. Nature is more disheveled. Even in controlled conditions, such as for fruits flies in laboratory test tubes, populations do not always follow the exact growth curve. In nature there are predators, competitors from other species, variation in resources over time and space, and unexpected disturbances that can push or jolt a population off the assigned trajectory. Yet breathing at a millennium pace will allow a glimpse of these dynamics. They may be noisier than the theory would predict but they are the engines that result in succession.

One thousand years is a long time, lets say 13 human generations. Go back 1,000 years and you could have a chat with your great-great-great-great-great-great-great-great-great-great-great-grandfather. As an old man he might have been around when Pope Urban II calls for a holy war to wrest control of Jerusalem from Muslims, a decision that launches the First Crusade in 1096. This kind of timescale is beyond our ability to intuit. A minute of breathing and we have moved 12,000 years, which if we turn the clock back takes us way beyond recorded history. And yet it still sits within ecological time and only starts to tickle at evolutionary time for organisms with short generations.

If we let the clock run backwards and breathe for a quarter of an hour we will be at the point in evolution where *Homo sapiens* first appeared. Now we are into evolutionary time. It will be 3.5 hours before we get to the first ancestor in the genus *Homo*, another 3.6 days before we get to the first primate and nearly 10 days more to get back far enough to see the first mammal.

The fossil record demonstrates that species go extinct and from analysis of extinctions it is possible to estimate the 'lifetime' of species. Insect and other invertebrate species seem to manage between 5 and 10 million years, whilst mammals are around for between 1 and 2 million years or a couple of hours worth of millennium breathing. It is a struggle to comprehend evolutionary time.

All we need to know here is that there are processes operating at rates that we do not see when we function in our normal time; and that some of these processes we have trouble recording, even with the diligent application of modern science. Things that are difficult to see we find hard to comprehend.

It is little wonder that when we press the reverse button and return ourselves to breathing in real time at a dozen breaths per minute, with the trees swaying gently and the flies buzzing once again in our ears, most ecological processes disappear.

The human experiment with nature

Now that we are back in real time we can return to the Balinese countryside around Gunung Batur. If you visit Bali you will find the usual bustle and closeness of Asian countries and an economy that relies heavily on tourism. What you will also see is an agricultural landscape. The streams that dissect the lower slopes of the volcanoes run through and provide the water for a patchwork of paddy fields. The clever layering of the terraces allows gravity to spread the water and still leave space between the fields to grow banana and coconuts palms. Livestock are kept intensively with the only free-range activity being by ducks. It is hard to find plants that are not providing some support to the people.

The 1917 volcanic eruption was not followed by a natural succession to tropical rainforest. The human population halted the process in order to appropriate all the primary production.

There are two things about this observation. First, that the appropriation is highly efficient and complete. Not a square foot of southern Bali seems to be following a 'natural' succession; even the forest reserves, that for cultural reasons are set-aside for monkeys, are managed through plantings. Second, that this change to the landscape happened, in the context of our discussion of time, very quickly.

On ecological timescales humans alter landscapes in an instant.

There is no reason to single out the Balinese. Human appropriation of natural resources has, in a flash, altered landscapes the world over. For example, in 1788 Europeans arrived in Australia with the intent to stay. At that time perhaps a third of the continent had forest cover; in 2003 cover stood at 21%, a reduction of over 900,000 km^2 or slightly less than the land area of Venezuela and twice the land area of Sweden. Around 4.5 million km^2 (58%) of the Australian continent is under some form of agricultural land use. The global average is 40%. Everywhere has been affected.

The rate of human expansion and alteration to landscapes since the industrial revolution is far faster than many ecological processes. Modern humans have cut and cleared forest, appropriated water and plundered wild populations for food at rates far in excess of nature's relaxed pace. Landscape change is therefore a disturbance event in the ecological scheme of things.

If we were breathing at one breath to a millennium the change would have happened in the pause between inhale and exhale.

So far we have very little idea of the consequences of such a large-scale disturbance. Although a similar one in the distant past resulted in the extinction of the dinosaurs.

Chapter 9

Essay

What if it's not emissions?

There are spectacular cliff faces in the Blue Mountains west of Sydney, Australia that are very old. They were made when sandstone sediments, uplifted 170 million years ago, became dissected by erosion. They have been wet, cold and hot more times than we can imagine and been the backdrop for millions of rainbows. If these cliffs were able, they would remember times when it was generally warmer than now, also when it was colder, wetter, drier, windier and times with more frosts. They would remember as though it were yesterday the last glacial maximum that began 20,000 years ago, persisted for several thousand years, and saw glaciers appear in New Guinea. Climate change for a rock is a bit like the weather for us.

Climate change is real. It always has been. All the scientific evidence points to it. Temperatures recorded on the ground, in weather balloons, and by satellites, as well as scientific inferences from radioisotope analyses in ice cores show change. Pollen in mud cores from old lakes shows changes over time in plant species with temperature preferences. Then there is evidence from the longer timeframe of stratigraphy and our understanding of evolution that suggests wetter, drier, warmer and colder conditions than the present in times past. All these tell us that our climate changes over time.

The only difference between climate and weather is extent. Weather happens locally and quickly compared to climate that covers more area and changes more slowly. In the time frames of days, months and years that we understand well, climate is more predictable than weather. In any one place on the globe the weather conditions will change daily and with season. Climate will change over decades when it is really in a hurry, but usually over millennia or an even more leisurely pace. So whilst we might like to think that climate is stable, it just depends on how you look at it.

We have already described nature pulsing to its own rhythms as it clings to the surface of a dynamic planet - a molten ball, with the thinnest crust, wobbling around a space orbit. We tend to imagine life clinging, precarious, on such a thin band of habitable conditions that seem to be ever changing. The reality is that life holds on rather well and in profusion. Ever since it got going, life has been very happy indeed with the conditions on earth, so much so that it has persisted for billions of years in spite of the vagaries. Organisms have wriggled, crawled, run, walked, flown, burrowed, diffused or drifted on the wind into every crevasse, crack and pore space from the tallest mountain to the deepest ocean trench and everywhere in between. Life loves it here.

Indeed, the constant fluctuation is why there is so much life. Change in environmental conditions is a key driver of evolution. For example, the leading hypothesis for the extraordinary biodiversity of forest guenons[24] in the rainforests of west Africa - there are 25 species and 34 subspecies - is a result of cyclical climate change. During cool times the forest was reduced to isolated fragments that allowed allopatric speciation.[25] If the world were static there would only be a fraction of the current biodiversity, our ecological processes would be run by fewer species and a very different set of natural resources and resource flows would be available for human use. And current theory predicts that less diverse natural systems would be at a higher risk of collapse.

Change in climate is just a part of this long-term motion. A change that nature thrives on, that fuels the process of

---

[24] A group of primates belonging to the genus *Cercopithecus*
[25] Evolution as a result of geographic isolation; the most common way that new species arise.

evolution and has helped deliver the many wonders that we have the cognitive abilities to perceive.

The odd thing is that although all the evidence points to a planet in constant flux, our human preference is for stasis. We are uncomfortable with change of any sort, including that of the climate. Uncertainty seems to freak us out more than anything else. Although crossing a busy highway is far more dangerous than a swim in a shark infested lagoon[26], most people would choose the former because we prefer it when we can see the danger coming.

Fortunes are spent to mitigate unpredictable risk in an attempt to buy some peace of mind. Just ask your insurance broker after feeding him a few beers. So keen is the human desire for the certainty that we believe is born of stability, that we have invented any number of ways to try and achieve it. Collectively we call it engineering - the application of science to the design, building and use of a mechanical contrivance. The engines are set to work to modify our immediate environment to help maintain the conditions and resource production that we desire.

Changing climate disrupts our sense of certainty. We would feel better if the magnitude and intensity of fluctuations in rainfall, temperature and storm events were known. If rainfall were regular, not too much or too little; frosts restricted to winter and never pitch up uninvited in early autumn before the crop is in; and temperature a balmy 25 degrees; then our farm businesses would be so much easier to run. We could invest in equipment, fertilizers and other inputs with confidence. We could even leverage our predictable production to secure loans for all kinds of things not related to production. Output would be guaranteed with flow on effects to the stability of markets, price and profit.

Even if the Gaia hypothesis[27] is true and the earth is an organism capable of regulating herself (somehow the earth has

---

[26] In Australia there are 1,600 road fatalities every year but there have been only 61 deaths from shark attacks over the 50 years to 2004, or 1.2 per year.

[27] The Gaia hypothesis proposes that the living and nonliving parts of the earth are viewed as a complex interacting system that can be

to be a she), feedback mechanisms on processes are never perfect, not even in the best of engines. There will always be some over and under shoot before any mechanism is able to stabilize a process sufficient to effect constant conditions. There are limits to how stable things can be even if Mother Nature is watching over us. Process and event cycles that occur over decades and centuries, the ones we have difficulty perceiving, are slow to catch up.

On a global scale, the huge temperature sinks of the oceans and complex dynamics of the atmosphere work slowly. They preclude our normal approach to obtaining homeostasis, which is to let technology deliver rapid engineering solutions. It is one thing to burn fossil fuels in vehicles around the world every day[28], another to assume that we do, or can, modify the intricacies of global circulation systems to our advantage.

We would be presumptuous, even arrogant, to think that humans can stop climate change. We cannot, any more than King Canute could prevent himself from getting wet as he sat on his throne in defiance of the advancing tide. It is possible for us to admit that climate will change. And it is possible for us to mitigate against climate change effects on our production systems. But to stop or control climate change is not in our power.

Reluctantly, and with a nervous sigh, we begin to accept that there will be some change. The earth is dynamic. We know this. But if change must occur, then at least let it happen in a predictable way.

This is our next most comfortable state after stasis. It is also why every media channel has a weatherman. We would like to know, or at least have some idea, of what will happen next. Consequently a capability to predict climate change has become something of a quest.

Driven by our desire to know we have asked and sent our best minds in search of answers to questions such as: What is the cause of climate change? Is there one or are there multiple causes? Are we, or more strictly, are our actions, a new

---

thought of as a single organism capable of self-regulation. See Lovelock, James (2000). *Gaia: A New Look at Life on Earth.* ISBN 0-19-286218-9

[28] In 2004 there were 13.5 million vehicles in Australia that traveled 200 million km and burnt 27,500 million litres of fuel.

cause of climate change? What will happen to climate globally and locally? We have asked scientists to come up with answers to these climate questions. In the US alone, we have asked for at least US$15 billion[29] worth of answer since the year 2000.

The majority view of the scientists, including those who have spent this money, is that:

- human activities have affected climate; specifically that the amount and rate of fossil fuel consumption has influenced the greenhouse effect
- global average temperatures will rise and change the patterns, frequency and severity of local weather events just about everywhere
- as temperature rises there will be consequences for global atmospheric and ocean circulation, sea levels, frequency and severity of regional drought and cataclysmic floods
- the consequences of the feedback loops are uncertain

This view from science is fast becoming the conventional wisdom. Whilst the causal link between human activity and climate change remains the most contentious, the empirical evidence has convinced us. We accept that the planet is entering a period of climate instability.

However, there are scientists (and not just those who are funded by the oil cartels) who are not convinced that our burning of fossil fuels is the cause of climate change. They either question the data or quote times past when the concentration of carbon dioxide was many times greater than at present. They note with a smile that at these times the globe did not cook. What happened was that, after a geological while, conditions emerged in which humans evolved. They say that climate just changes and that human influence is trivial in the dynamics of a world more than 4.5 billion years old.

Rather than totally ignore these scientific skeptics - no relation to our friend from the pub - there is a place for at least one valid point that they make. It is that the earth has had very

---

[29] US$15 billion is roughly the GDP of Ethiopia or Bahrain and there are more than 100 countries with a lower GDP. At the other end of the scale US$15 billion is around 2% of Australia's GDP and the Australian military budget for 2005 was US$13 billion.

different atmospheric composition and climatic regimes in times past. Not just the cooler times of the last glacial maximum, but times when carbon dioxide was more concentrated than today, times when ice may have covered the entire globe and times when Greenland was indeed green.

In the end it is trite to argue the scientific case for or against climate change, even for or against human induced climate change, because climate changes. This much should be a given.

There is another thing that some of the climate skeptics, a kind of collective noun for naysayers, have pointed out. The reality is that we cannot do an experiment to test the human induced climate change hypotheses. This means that deductive reasoning, the biggest and most powerful tool in the toolbox of science, is not available to answer the questions posed about climate change.

As applied by scientists, the process of deductive reasoning develops arguments to move from given statements to conclusions that must be true if the statements are true. In science, this deductive process is built on the results of experimentation designed to provide evidence for both the statements and argument and so provide rigor to the conclusions. This evidence-based reasoning is the foundation of our scientific knowledge and its spinoffs in modern technology, health and societal systems, but requires that realistic controlled experiments can be conducted. However, there is only one earth available to us. So for questions of climate change, we cannot produce the conditions of replication and controlled manipulation that characterize experiments.

Given the egotistical panache of scientists, the mere hint of uncertainty can get us into a mess. Everyone becomes embroiled in an 'I said, she said, he said' pseudo-scientific debate between the believers and the skeptics – an example of which was aired recently on the Australian national broadcaster, the ABC. A televised debate became a slanging match over whether climate actually changes. Both sides were correct but the time lenses were different. What we can say is that without its key weapon of experimentation, science cannot really help us decide.

It is possible to use deductive reasoning to understand the local effects of climate on processes – such has how a wheat crop might respond to elevated atmospheric CO_2 – but we cannot run controlled experiments on planetary circulation. All that we can do is observe and then induce, *post hoc*, from our observations. We are forced to rely on inductive reasoning[30] and this carries uncertainty.

In all areas of science we are cautious about this kind of inference. The conundrum is well illustrated by the problem of correlation where two variables can be closely related but the relationship does not imply cause. The size of a mans girth at age 50 correlates with his risk of heart disease, but it is not his beer belly that clogs his arteries, a sticky substance in his blood does that. It is difficult to experiment on the causes of heart disease due to the absence of test subjects yet your doctor is very cautious if you are over 50, of ample girth and present with symptoms of indigestion and lethargy.

In the scientific debate around climate change we have relied on models to make predictions. Models are built on assumptions and so their predictions will be, by definition, only an estimate of reality. Models give us an idea of what might happen. They give us something against which to test reality or to choose among so that we can prepare for outcomes that the models predict. The problem is that models are based on inductive reasoning too. They are only as objective as the assumptions used to build them. Even the best of them will not include all the possible feedbacks, threshold effects, nonlinearities, random effects and other sources of variation that deductive reasoning tries to capture and explain.

Models are very useful abstractions, but it is easy to debunk them because inductive science comes with uncertainty as sure as burgers come with fries.

This digression teaches us that if we wait for the scientific debate to give us a definite answer to our climate

---

[30] The other major logic tool available to scientists is inductive reasoning, which reasons from a large number of particular examples to a general rule. The application in science involves the statistical analysis of pattern or trend in data to allow a conclusion(s) to be drawn. All computer models, including those used by members of the IPCC to draw their conclusions on climate change, are based on inductive reasoning.

questions, it will be a long one. Climate change will be happening while we fiddled. This fuels those most fearful of change. They say that there may be some uncertainty but we cannot wait around to be more certain. The scenarios that the models describe are dramatic enough for us to act now. At their core, this is the message of the IPCC reports[31]. Perhaps it is this feeling of a need to act that has captured the public imagination.

Despite being screened in only a handful of independent cinemas Al Gore's *An Inconvenient Truth* was watched by many and talked about by most Australians. The buzz was loud. It seemed to define a change in attitude sufficient for even hardened political skeptics in Australia to alter their stance on climate change. And there has been some action. More and more people are turning off lights, having shorter showers, printing on both sides of the paper and riding the bus to work. Earth hour was such a success in Sydney that it has been exported to other cities around the world. Suddenly public awareness has latched onto climate change and, in particular, carbon emissions. It would seem that we are becoming aware of our carbon footprint, the amount of CO_2 released to fuel our modern lifestyles[32].

Some commentators have suggested that the tipping point is past and the momentum is enough to create real action. There are some numbers to back up these hunches. Market research companies tell us that 70% of consumers say they are willing to do something that will help combat climate change. This is a significant proportion. However, there is a difference between what people say they will do and their actions. The experts call this behavioural inertia. Consumers hesitate until they understand what is being asked of them. But once people find out what to do, and it is made available, they will do it. As

---

[31] The Intergovernmental Panel on Climate Change was established in 1988 by the United Nations Environment Program and the World Meteorological Organisation. Its role would be to interpret scientific, technical and socio-economic information relevant to understanding the scientific basis of risk of human-induced climate change, its potential impacts and options for adaptation and mitigation.

[32] In 2006, average CO_2 emission in Australia was 25.6, the US 24.5, Sweden 5.0 and India 1.9 tons/capita/annum.

we saw earlier, we will let complete strangers change our light bulbs.

What also happens with shifts in public opinion is that they have an immediate effect on business. Companies, and especially entrepreneurs, notice that consumers have made the link between carbon emissions and climate change. And they know that once this happens consumers are ready to adopt solutions to reduce their emissions. If there is no detectable effect on their comfort, wallet or standard of living, consumers will respond to the carbon neutral sales pitch. They make carbon friendly purchases because it increases their sense of well-being. Indeed it is likely that consumers will buy what they believe to be carbon friendly products before they take on direct costs such as installing a solar water heater.

Once this shift in purchasing behaviour is large enough, it creates a major business opportunity. Many, if not all, businesses can increase sales of products or services if they demonstrate green credentials. And there are opportunity costs to businesses that fail to demonstrate that they are sensitive to size of their carbon emissions. At the time of writing many businesses in Australia are starting to look at their carbon footprint, not because they are green, but because there is a business imperative.

The sequence they are following is: measure – reduce – offset. First find out what the footprint looks like in your products, operations and supply chain. Answer the question of how much CO_2 is emitted in the process of putting the product on the market. Second implement ways to reduce that footprint – turn off the lights, print on both sides of the paper, buy green power, purchase aircraft made of carbon fibre. Then if there are any emissions left, which, if your business is commercial aviation where 30% of your operating costs goes to purchase jet fuel, there will be. Then you offset that CO_2 production by purchasing carbon credits[33].

Businesses are already buying and selling carbon credits in advance of any national or globally ratified trading schemes.

---

[33] A carbon credit is a tradable permit that puts a value on greenhouse gas emissions. A credit gives the owner the right to emit one tonne of carbon dioxide.

The 7th Annual World Bank carbon market intelligence report determined that the global carbon trading market tripled in 2006 from US$10 billion to $30 billion. This official regulated market sits alongside a voluntary market of US$100 billion. Put the two markets together and you have a sum equivalent to 18% of Australia's GDP.

This tells us a couple of things. First, that carbon trading has been around for a while and second, that it has reached its takeoff point of exponential growth. Long before Australia has any official trading system, the market has created one. Corporate Australia is already very interested in carbon because there is money to be made. And for some companies the environmental department will go from a cost centre to source of income. In time the environmental manager might even get his own office.

All this came about because consumers perceived carbon emissions as bad. Business recognized and responded to the opportunity.

One outcome is a new market, created by giving carbon an agreed unit of measure (mass of CO_2 in tons) and, subsequently, a monetary value. In time the same will happen to water, biodiversity (bio-banking being a start) and eventually ecosystem services. Business has taken hold of the shift in public perception long before it was possible politically.

There is much discussion to be had about the implications of making carbon emissions into money. Debates will tackle the efficacy of the market mechanisms to seriously reduce emissions, the intricacies of a multitude of trading products, the relative and real value of any given credit, cap and trade schemes, pegging quantity or price, and many more increasingly detailed subtleties of the instrument. The important thing is that the debate has started. It is going on among people from the finance, insurance, investment and accounting sectors together with a growing number of service providers that have emerged to help the corporate world trade carbon. Money people are talking about carbon emissions.

What all this means is that the big end of town suddenly has a reason to be involved in the environment. For once the environment is not something that must be kept off the balance sheet at all costs. This is an extraordinary opportunity. Perhaps for the first time, business has a real reason to be informed by environmentalists and natural resource managers about the importance of the environment, and not just about the implications of the regulations on the rare and endangered. They have a reason to know something about how the environment works.

Those with a business eye to the future, together with those convinced that humans have caused the latest climate change, hope that the carbon market will help economies reach abatement and emission reduction targets. The politicians hope this will happen with bearable impact on economic growth and prosperity. So those with political leanings, green affiliations, and/or a belief in human capacity to reverse change have entered another debate around formal mechanisms to reduce emissions.

The new federal government that Australians elected to office in November 2007 had an Emission Pollution Reduction Scheme as part of its election mandate. The EPRS is a cap and trade scheme. The idea is to let the market facilitate a shift away from emission intensive means of energy generation and manufacturing production. The debate about who wins, what losses are tolerable and what compensation is necessary following the implementation of the scheme has begun to fill newspaper columns and press releases from all stakeholders.

The focus on the detail of winners and losers has overshadowed another debate. The important issue of whether emissions will be reduced or if reduction will happen fast enough[34].

At times it seems that politicians and business leaders - especially through the skills of industry lobbyists - have become besotted with the idea of trading carbon. Everyone seems to have a passion for either debating the issue of carbon emissions or trying their best to benefit from it. And all because

---

[34] The Stern Report, released on 30th October 2006 by economist Nicholas Stern for the British government as a summary of the effect of climate change and global warming on the world economy, predicts that if immediate action is taken to reach reduction targets the cost will be moderate (1% of global GDP) and global temperature would be expected to rise by 2.8 degrees. Failure to reach the targets or even a small delay in reaching them could see temperature rise by 5 degrees with a much greater cost.

somewhere along the way public consciousness was awakened first to the idea that climate change was coming and second to the notion that if we fixed up carbon emissions all would be well.

Let us return to the opening stanza of this chapter. Climate change is real. It always has been.

Based on all the evidence we have it is highly likely that the climate of planet earth will change. Trading carbon is an interesting departure that has taken our eye off this basic truth. Now we might ponder the title of this essay. What if it is not emissions?

Consider for a moment if the skeptics were right and human release of CO_2 was not the trigger for the current change in climate. Or if emissions did trigger the change, what if reducing emissions will not reverse or stabilize climate change. The long cycles and slow feedbacks we have talked about previously make instant stabilization from emission reduction seem very simplistic. What if it is not emissions? Well, we have been fiddling. Fiddling at what we have done for generations, which is to find new ways of making a dollar.

The response we have made to the public fear of climate change is to create a market for carbon. This has happened because we believe that this is the best way to reduce carbon emissions in a world that would collapse if its economic systems were not protected from undue disturbance. So far we have not addressed the whole issue. All we have done is to set up a system to make a handy buck from addressing the symptoms.

A carbon market will not, by itself, mitigate for climate change effects. When the climate changes come, we will need resilient landscapes, ecosystems and habitats that will still function to provide goods and services within a socio-economic context that works. We will need adaptation. The real debate should be whether the tendency of the planet to exist in different climatic states has import for us now. And when climate changes, what will it do to the environment that now supports 6.5 billion human beings scrambling for resources?

The good news has to be that sequestering more carbon into our terrestrial and aquatic ecosystems is good for biodiversity and ecosystem services. More carbon means more biological resources in the landscape. There is, for example, a lot of carbon credits to be had if Australian soils are improved form their denuded 1% organic matter back towards the 3 or 4% organic carbon they are capable of holding. Soil with more carbon holds more water, exchanges more nutrients and grows more plants, which in turn fix more carbon. It is a reversal of the cycle that has mined soils the world over. There are also some innovative ways to make this happen in Australia[35].

It is best if the scientists, practitioners and conservationists do not get embroiled too much in the carbon trading debates, even if it smells like trades are a license to pollute. These folk need to focus on the adaptation strategies. We will need them to develop the techniques that will ensure carbon is fixed and stays fixed in ways that deliver a complete range of environmental benefits. And because it is in the interests of the carbon markets, we can use some of the money that these markets generate to make it happen.

We are not far from new commercial options for natural resource use. In parts of the landscape there will be a greater, more reliable commercial return from growing trees not for their timber, but for their carbon credits, biodiversity credits and environmental service stewardship credits. We would finally be valuing the carbon in the trees, the carbon in the soils and the biodiversity that holds it all together. This type of management would be part of the mitigation against the climate change effects that a 3°C rise in temperature would bring.

Our challenge is that markets drain us as individuals. They are designed to provide us with money in exchange for our time, only for us to give it back in ever more imaginative ways. The effects of climate change are likely to drain us further. It is risky to allow 'the market' to solve the climate change problem that we perceived because the market is designed to make money. It is not there to encourage practices that increase soil cover, raise soil carbon, have better on site nutrient cycling, improve landscape connectivity, integrate catchment scale management or the plethora of other foundations for climate change adaptation.

---

[35] An open minded read of Peter Andrews "Back from the Brink" may be rewarding

Chapter 10

Story

Ants

I am lying face down. It has been dry for weeks and there is a warm sun on my back. The ground slopes away to a creek. This makes lying down more acceptable, even if prone on mown grass smacks of falling to claim a foul from the referee.

I had to get out. Snatch some sanity from the jaws of chaos. I had to connect somehow with something that matters.

Rows of maple saplings have been planted. I guess to make the place more like a park. Maples are not native but they will look good in autumn and will generate shade in fifty years time. I will be gone by then. I will be gone but I keep going back. Why I keep going back every day amazes me. Why? Back to do the same things over and over. I am crazy, or I will be.

At least I can feel the grass under my fingers. It is another alien species, not a native of course, because lawns rarely are natural. I once read that hippos create grazing lawns, but I don't think it's quite the same thing. I believe this is a couch grass that I can feel feathering this outdoor bed. Turf companies sell it but it's also a weed. Funny that.

There are not enough runners to be kikuyu. That one has every blade the same, clipped square at the top by the mower to leave a brown crown. Wait, this is different, a *Poa* or some other thin leaved thing hanging on in the couch forest. Whatever this species mix, it is more real than I am.

My arms hurt. Soon I will need to get up and move around. Pity those saplings. They are going to stay still for fifty years. They have more patience than I can muster, no, more than I can imagine. Maybe I'll stay down and suffer through it, yes I will. Resting on bony elbows eases my view of the world.

I can see a bare patch of soil beneath the nearest maple, just waiting for the grass to grow back if tree allows it. What does that mean, 'if the tree allows it'? As if it had any say in the matter. As though it can proclaim its space, then phone the garden police when the weeds, like any true delinquents, pay no attention.

Would you believe it? There is a sprinkler head between my elbow and the bare patch. Is there any part of this planet humans haven't altered? I doubt it. I remember once, on an expedition in the outback, far from any real road, trekking across peculiar expanses of soil with stones on top they call gibber plains, we would stop when the machine went beep to take measurements and there, in the middle of nowhere, there was always some evidence of humans. Every time there was something human lying proud in the dirt. It was amazing. I mean it was amazing to see a fence dropper or pieces of wire or spent cartridges or tin cans or a soda bottle from the 1950s or some other cast away, in the middle of a gibber plain. How does litter spread so far in just 200 years? It means that there is no nowhere. Somewhere is everywhere because a man in a tattered felt hat forgot to pick up his mess.

Amazing. An ant just grabbed a piece of my chicken and avocado roll. Not a big piece, just a crumb, but it must be at least as heavy as the ant given it's a bit of crust twice its size. It's an *Iridomyrmex* I think, the ant that is, a hyperactive critter that moves like it has some mental disorder. If I moved around in such a fashion the G-forces would tire me out in minutes. Luckily for the ant, it has a tiny mass that generates little momentum and it is a strong bugger. In an instant it has made haste with the crumb from my roll.

I stare, no longer aware of my elbows, and I see he has friends. They seem to like the surface of the dead leaves – ant flyovers that give them room to move amongst the skyscrapers of couch grass. It must be like running through a forest of trees. Just like the scooter scene in the Star Wars movie.

Maybe they are not that crazy. They seem organized. They know how to follow a system of trails and greetings that lead back home, which for this mob is somewhere beyond the sprinkler head. I follow one guy - somehow it has to be a guy - who, with quivering antennae that brush and interlock, talks to six other ants in the space of a few seconds. Time enough to cover a couple of dozen ant body lengths. I never talk to everyone I meet. I try to smile. And I try to say hello. On good days I manage it, but in a crowd there are too many folk, too many hellos to make. My antennae fail to quiver.

I go back to the ant carrying his piece of my bread roll. He has already managed to move it, in effortless motion, nearly a metre. My mental arithmetic figures out how far this would be if it were me. I would be carrying a cardboard box, bulky as a television, from my grassy lounge back to the office...quickly. And even if I were weary in well doing, I wouldn't be allowed to break a sweat because the ant is effortless. Lucky for me that Newtonian physics makes my comparison dumb. You have to agree that ant fortitude is impressive.

His mates have kept up their endless hellos and passed on the message. More of my lunch starts to move.

The book says these *Iridomyrmex* are native. It says that they evolved here; well their ancestors did at least. They started amongst the eucalyptus but it seems no matter to them that their patches of the world are now couch and maple. I can't see that they mind. If they recognize bread roll as edible then they must have Catholic tastes or maybe a sense of reckless adventure. I never thought of Catholicism as a useful survival tactic, but then I have never been reckless. Maybe I should have had more adventure. It might have helped.

Then again I am a newcomer. This place is new to me. Old hands tell me that fifty years ago this land with lawn and Maples was paddocks and market gardens. If the spot where my elbows now scream was a paddock, then I guess a generation or two back an *Iridomyrmex* colony would have had foragers carrying off thistle seeds and unfortunate pasture bugs. No matter to them that the incisors of horses and goats mowed the grass. With no shade trees it must have been pretty damn hot in summer. That, I imagine, suited them just fine.

I will be dead. Fifty years from now is too far for me to think, but I can see this mob still here. Fifty years of constant scurrying for edible scraps. Not a care over natural or introduced. These ants may be hyper, but there is efficiency in what they do. It's admirable.

I strive, at least as hard I think, but there is less joy in it. I wish I had their simplicity of purpose. If only macro-life was as easy as this tiny ant world. Maybe we have physics to blame. Hang on. Fifty years ago, my parents hadn't even imagined a twinkle. I was still a decade from conception. Somehow that means nothing to anything.

The ant with his part of my lunch has reached a small hole in the ground. At a guess the breadcrumb is roughly the same size as the whole. He waits. Five, no six, of his nest mates emerge, but no hellos this time because the crumb gets in the way. And anyway, the emerged seem to be on a mission. They scurry off at a pace.

There is no signal that I can see but my ant decides to plunge into the hole. Sure enough, the crumb is the same diameter and gets stuck. A few exploratory tugs where the prize starts to look like a bath plug, and then it's gone as a Herculean pull consigns it to the depths. In an instant another ant emerges from the hole. Even with my nose to the ground all ants look the same, but it didn't seem like there was enough time to change shifts, so I cannot tell if it is my ant.

My elbows finally give out and I roll over. It is blue above, just like it was fifty years ago. Any clear day to come will be as blue. That, at least, is reassuring. Will I lose the job that drags me back to my dreary office? My gut tightens because it appears so temporary. I mean it wasn't around fifty years ago. No one measured biodiversity then. The idea hadn't been born. Even its older sibling ecology was barely out of diapers in the 40's. Although I wasn't there, nor have I had the gumption to ask any of the old ones who were, I know that the sky then was blue. It was just as blue as it is today. It was blue.

Ants would know that the sky was blue. They are timeless. Haste without hurry as they forage for scraps, only able to discern the difference between edible and the rest, so they would have time to see that the sky was blue. I would like it to be that simple. I would like to have only simple, easy, uncomplicated decisions. Shit! No decisions at all, just instinct. Yet I am cursed with decisions, choices that I must make. Then I am going to die. The only decision I would like, but cannot make. Not my choice.

I wonder how old my ant is. He has probably been scurrying for months, or maybe he is new to the profession. It is impossible to tell and I doubt if he either knows or cares. He was born with all the skills he needed and his career mapped out. Instinct covered all his bases from day one. No women to chase, babies to raise, mortgages to pay, doctors to see, just the daily round of haste towards anything edible. Career decisions loomed over me before I had even a handful of life skills and have been perched there ever since; merciless as they suppressed my instincts. No worries they say around here. If only.

I guess the ant colony could easily be older than I am.

No one ant gets that much time, but the genetic integrity of the collective certainly does. There is a bigger individual. They can even share the responsibility to procreate, lucky buggers.

I roll over again and soon locate my ant on another expedition. I know it is my ant, otherwise why would I have rolled over from my blue view at exactly that moment? Fate of course. It is the way of all things.

He is following the exact trail that leads to the remains of my roll. I don't believe in chemical trials, bunkum. I reckon he just remembers where he found the last piece, just like desert ants that can navigate from mental images of star systems in a pinhead of a head. Now that is nanotechnology for you. I watch him follow the exact route, past the same few stalks of *Poa* that must look to him like bamboo amongst eucalypts would to me. Then no matter how he got there, he found another crumb, just where the first one fell. It is a bonanza, a fantastic feast that is worth telling your relatives about. It would seem that he has told them, as there are many more now than before. Somehow all that antennal touching fuses unbreakable bonds amongst kin. Maybe that is chemistry. It creates a loyalty that knows no bounds, generosity to share and protect without any hesitation.

There he goes, touching again. That's three greetings in as many seconds. He must greet more folk than the Pope, and they're all relatives, close relatives at that. All I remember was intense rivalry with my brother. He was on a mission to appropriate as much of the attention as he could and he was very good at it. I was sure he played up just to hone his skills of resource gathering, because he never seemed to really need all that attention. My ant seems only to behave for the collective good. He is a true egalitarian.

Surely there has to be something in it for him.

I am forced to compete. Not with my brother anymore, he is on the other side of the world trying to get people to buy his clothes - well, his company's clothes, naturally – while I am trying to make a living selling science. So I don't compete with him anymore. At least I assume that I don't, unless I missed something. It would be nice if when I saw a contract or a customer I could say 'hey guys, we have a customer over here, come lets join in', and we would all get together to provide the service quickly, efficiently without squabbling. But that is not how it works. Customers for environment services are like hens' teeth. This means they are very hard to see and small. So when one is spotted it is a mad scramble like lions around a kill. Yes, surprisingly messy, bitchy and cat like as there is no love lost at the pride dinner table.

Ants don't squabble. Theirs is a very organized scramble. The only hurry is to get the resources below ground before another species, so your own kind is roped in quick smart to get the job done.

There are six, no ten, at the crumbs now. Lucky for them it was crusty bread.

I will not be here in fifty years time, but the ants will be. As the line keeps going, nephews and great nephews will follow these ant brothers for a long time. If the colony dies, the space will be taken up by relatives, maybe even direct descendents, sons and daughters that flew away on a sultry summers evening. It must be reassuring. I am going to die without any idea of my space, let alone if my descendents will be around to fill it. I don't even know if I will have any descendents. My daughter is at high school carrying around a complement of my genes but she is not remotely interested in procreation. A commune or kibbutz's is more her style, and whatever happens she has wanderlust.

My ant has found another crumb. Mercifully it is half the size of his last trophy, yet he carries it in the same purposeful manner, high above his head, effortless. He is a worker for sure. No sons or daughters only brothers after a fashion, yet constantly striving to provision the next generation with selflessness for the common good. Surely he has thoughts of escape. To wander off like my daughter will. Seek new ground, new cuisine and ways of living.

I would like to escape. Let my mind sip the freedom of instinct. Act only on what feels right. No longer to be chained down by a set of rules I had no say in making. Yet I am shut in a prison where the guards are adamant in their chant: no work, no food. Maybe it is the same for my ant. Maybe he feels the constraints of a foraging life. Collect crumbs or you don't eat.

No, I don't think so. He cannot operate like that. He is free. Please god, if you are there, let him be free.

I remember one time there was a bullfrog on my back lawn. It was wallowing in a bonanza of food that even frogs must dream about. All the local colonies of white ants had sent their reproductives on the nuptial flight. There he was, this cumbersome beast, idle on the lawn picking off alates with his sticky tongue until he had eaten so many that his stomach, gullet and mouth were full with wriggling termites. Every time he caught another, three would escape. He was satiated, but he couldn't stop the reflex to feed.

I am like that. I am full, but reflex sends me out to compete, to do my job every day. Even when things escape every time I open my mouth. I am sure my ant forages for the love of it. He walks to the food for the sheer joy of the search, thrill of the find and the satisfaction of carrying the prize home. Hey, not only that, he gets to do it with his relatives, argument free.

I must come back as an ant. Finish this human existence that forces change and destruction, come back as an ant to tidy everything up.

I wake up. In the ungainly effort to get onto my feet I step on the hole that swallowed the crumbs of my lunch. There are still ants returning with provisions and leaving on the next mission. They will be here for a long time.

I gather myself and pick up the paper bag that kept my roll fresh. I carry it and my malignant tumor back to the office. It is very heavy.

Chapter 11

Story

Little Creatures

The heat had been building for a month but Tebogo had not noticed. There were chores enough to fill the daylight around her daily trek to school and her siblings were always a handful.

A breeze lifted a little of the silt from the sand around her feet as she brushed at the tiny acacia leaves that her mother wanted gone. This day there was no school so Tebogo's mother had taken her brother and sister to see their aunt who lived in town. It took many hours to make the return trip and Tebogo had been asked to mind the vegetables. The family spinach was in a precarious state because Mma Raseroka had been letting her goats run unattended through the village. Spinach that was almost ready to harvest was irresistible to a goat. Tebogo was glad that her parents would leave nothing to chance, especially with errant goats. It meant that she, the eldest and most responsible one, would stay at home to mind the spinach.

This was a fine arrangement. Tebogo did not care much for her aunt who spent all her waking hours with her jaw in motion. When the padded bone moved out would come an opinion or half-truth about someone or something. Relatives usually, but neighbours were included too; even passers by should they be unfortunate enough to catch her glance.

Obviously the young were not supposed to listen to such things. Tebogo tried but always found that even when she shut her ears, the noise of gossip left something tacky on her. It was irritating and not something she liked much. So she was happy to stay behind. It was good to be home on a warm day.

A posse of young boys passed by the gate, chattering like monkeys. They looked across and jeered a little in her direction. Tebogo was too young to be bothered by boys, although the changes she could feel made her think that very soon she would interest them greatly. For now she could push such dangers aside and focus on how lucky she was to be the eldest and old enough to boss the little kids. Mother said that Tebogo was to watch over and play with her sister and her baby brother. "Be nice to them," her mother said. But Tebogo knew they would have to listen to her when she was a woman.

Today there was no sister with a snotty face and no baby brother squatting in the dirt. All that Tebogo had to do was to watch the goats.

There were two important things that everyone knew about goats. The first was that they tasted good, especially on feast days; and the second was that they were always moving, except on feast days.

They move around all the time because they think that there is going to be a better mouthful somewhere else. They think that the next bite on the next branch, or the next bush, or across the river, will be sweeter and juicier than the last. They just have to keep trying to find the most delectable leaf there is. Along the way, on this never ending search for the juiciest and sweetest, they eat everything.

Spinach, of course, cries out 'halt' to any goat. On the mere sniff of a spinach leaf, a goat will forget all future hope, stop and eat this delicacy down to the stork.

Tebogo knew these goat rules because she had watched goats for a long time. She watched them wander past her family garden early in the morning without even turning a muzzle until the spinach leaves began to peak out from below the cattle droppings on the raised beds. If father had forgotten to close the gate they would march in with only one thought, and if he remembered, they would gather beside the fence until one of them could squeeze through the gap that her sister sometimes crawled through. It was then that Tebogo had to step in and irritate them with waves of the yard broom. This forced them to move on and to forget about the spinach until they made their rounds again.

There was another thing that Tebogo knew. Once the goats had been moved on there would be time. And she had a secret. She knew how to make this time perfect. Of course everyone knew that there was time, but not everyone knew when it was perfect.

"Time is perfect," Tebogo would say to her brother as he sat on her knee. "Time is perfect when you don't notice it."

The little boy would gurgle his agreement.

Tebogo would continue. "Yes, time is perfect when it slips by in no obvious hurry to meet or catch up with itself."

Gurgle.

Tebogo had always known exactly how to get time to pass in this way. And even when she had to keep an eye on her siblings, she could dream enough for the moment to last forever.

Her friends from school had forgotten. They always spent their time with slumped shoulders and sour looks, complaining that there was nothing to do in the village. They said that they were bored. All they wanted was to reach the future when they could visit town. Then they wanted the best future when they could move there to be with the boys.

Sethunye, who was Tebogo's best friend, already had a boyfriend called Mothusi. Sethunye said that he had a silver sports car with black leather seats and that this car could go very fast. All the girls wanted to see it but Sethunye said that the car wasn't fast enough to reach the village through the sand. This news did not stop the girls gathering at the edge of the village just to wait for the car to come.

The goats arrived. Their eyes bulged and their step quickened as they caught the bright green sails of the spinach plants that propelled the garden bed merrily along a sea of swept sand.

Tebogo raised her broom at the leading goat. Either a hundred stems of thatching grass tightly bound to one end of a mopane pole is very frightening to a goat, or it is easy to remember. The goat, a bold individual with a brown patch on its side, saw the broom and forgot the spinach. Spinach was certainly juicy and sweet but not worth a clip from that yard brush. As it passed the pumpkin bed the goat turned its head toward the broad leaves with longing but Tebogo waved the brush above her head.

The goat departed, resigned to breakfast with its cousins on boscia bushes.

Tebogo sat down on a wooden stool to watch the rears of a dozen goats pass. There were always so many. They made great stew, and the milk was tasty, but they were also plenty of trouble. When they wandered out of the village to browse they had to be watched in case a leopard might snatch one. Then at night they had to be rounded up into the kraal to keep them safe from lions and hyenas. At least that job kept those nasty boys busy. Maybe that was it. There were many goats to keep the boys from hanging around. How clever.

Lions and hyenas were both very fond of goat. It was good that a goat had a small brain and a short memory for otherwise it would die of fright when the lions roared.

Tebogo was not frightened. She often went with her mother to collect thatching grass in the game reserve nearby. Her mother would make them sneak under the fence and walk down towards the open grassland and mutter "Watch for that lion. See if he comes." Her mother was always afraid that the lion would come, or worse, the warden. Tebogo hoped that they would see a lion but her mother would make her chant as she walked to keep them away. One day a man from the Wildlife Department came to Tebogo's school and told all the children that collecting thatch and fruits in the reserve was a good thing. He said it was called "multiple use" and, so long as the villagers kept their goats and cattle on the village side of the fence it was ok to go in and collect grass. Tebogo ran home after school to tell her mother but still they went into the reserve at dawn under the fence and hid the thatch grass behind the maize store.

When she was not hoping for a lion, Tebogo watched the grass beneath her feet. The thing was that the plants in the reserve always looked better than the same plants near the village. Even in early summer when there had been no rain for months and the days were very hot, plants looked better without goats.

Now the goats were gone. Tebogo put the brush down against the fence and began to walk beyond her father's kraal in the direction of a stand of trees that the white men called apple-ring acacias because the pods looked like apple peel. Tebogo had seen apples in town, one time she even ate one that was very crisp but not as sweet as marula fruit, but she did not know what an apple ring was. Remembering the idea of a ring of apples she reached the far side of the trees, out of sight of the village, and jumped onto her favourite perch on a mopane tree. Elephants, especially those frustrated young ones, liked to push over mopane trees. The trees didn't seem to mind and usually just kept on growing their heart shaped leaves. Tebogo twisted her spindly limbs into a nook that formed when her tree recovered from past elephant attention. It used to be a perfect seat but lately it had become a tight squeeze. Now all she needed was a minute or two to let her breath settle.

At first she watched the sky. There were clouds scudding across the blue. Tebogo saw each cloud as a trotting eland, galloping zebra or a charging rhino. On days when there was only blue she would stare at the colour until a vulture or an eagle strayed from the reserve so high up that it was only a speck. She would follow the flight as the bird flew across the sky, always more smoothly than the clouds, until perspective allowed it to reach the line of trees on the horizon. Then she would seek other birds that lived closer to the ground until she found one on its way towards her. Sometimes it was a noisy babbler or a black shrike with the longest of tails, or sometimes a barbet with a yellow bill and crested head. She let the bird bring her eyes to focus closer and closer to her own perch until she could see the bushes across the clearing and the grass beneath her.

A Bateleur eagle appeared from behind a zebra shaped cloud. It wobbled its way across her gaze, drawing it down just as a Lourie screamed its goo-waay from a nearby acacia. The Lourie took off, startled by a rustle in the grass below. A mongoose poked a twitching nose on slender muzzle through sand coloured stems of a tussock; alert eyes ringed by light brown fur that made his small face bigger. Satisfied that vigilance was done, agile limbs pushed the mongoose forward. It twittered and began to scratch at the earth.

Soon a whole clan scurried by on their search for food. Each furry individual sniffed and scraped at anything that might reveal an edible morsel, then stop to sit up and converse as though noise made them feel safe. A young one let his nose lead him. He sniffed his way to the foot of Tebogo's mopane tree, close enough for her to count the bands of coloured hair across his back and to see his whiskers twitch. Oblivious to his watcher, the mongoose sniffed at a cone of loose soil and with a sweep of a paw the cone, no bigger than her brother's fist, was spread flat. The mongoose rummaged through the debris and crunched at tiny insects that scurried in all directions. The feast lasted only a few seconds before the mongoose, unable to keep still, had moved on and out of sight beneath her.

Tebogo smiled. The furry creatures searched for food just like little goats, always moving to find whatever they could. And just like the goats they left a mess behind.

She looked at the spot where the sand cone had been. There was a tiny hole in the ground. After a moment a string of insects emerged from the blackness and scurried around in all directions, bumping into each other in their haste. If they twittered Tebogo could not hear it, although it looked like they were trying to understand and tell each other what had happened. Tebogo kept one eye on the hole.

More little creatures emerged. It was not the safest time to be out in the sunshine, but the little insects kept busy. Work in the fields always happened in the morning before the sun got too high. Midday was the time to relax. If you were an aunt there would be a shady spot to gossip, and if you were a young one, you would have to just sit and pretend not to listen.

Tebogo let her eyes drift from the hole towards movement near where the mongoose had first poked its nose. One of her tiny creatures was running back and forth, then stopped to climb a dry grass stem. It bit into the top of the stalk and cut of a section about half the length of Tebogo' smallest finger. It carefully picked up the grass cutting and scurried along in a strait line to another hole, only this time without a mound of soil. As Tebogo watched many insects emerged, some were large, others small, but they all did the same thing, returning to drag grass back down the hole.

Tebogo wondered what happened when they went below the ground. Did they drop the grass, turn around and go get another piece? Maybe there were many, each taking a turn to gather the dead grass. And what happened to the grass? The earth was solid. There was no place for grass. In the sorghum field it was hard to get the hoe into the ground, it took all her strength, yet these little creatures went down into it as easily as the goats moved between boscia bushes.

The sun slid along its curve in the sky to change the light. Tebogo eased herself down from the mopane tree and ambled home to her chores. She wondered if the creatures kept gathering while she chased the goats and washed the dishes. Did they scurry all day, maybe even all night? If they did they would collect a lot of grass, perhaps more than the cattle with their heavy feet and fat stomachs. Maybe they even eat the grass just like the cattle.

The next day at school she would ask her teacher, a very clever man who would know about the little creatures she had seen.

When Tebogo asked, her teacher just smiled.

"Those creatures are termites Tebogo." He said. "The white men who know them have named them *Hodotermes mossambicus*, and they say that they live all over Africa. The farmers call them harvesters and they do not like them. They say they eat too much grass. Did you see they were dark in colour these little insects?"

"Yes rra."

"This protects them against the sun so they can collect grass even during the heat of the day. And they are many. They say more than one million in one nest that can be as big as a boma."

Tebogo stared in wonder, amazed at how much her teacher knew about her creatures.

"The holes that you saw are connected under the soil so that the termites can store grass. They must have help from bacteria in their stomachs to get the goodness from those hard grass stems. No other animals can get this goodness; the termites can because they have bacteria to help them." The teacher smiled as Tebogo began to wander away, forgetting her manners.

How many is one million? People did not count that far. They counted one, two, ten, twenty, and then many. One million was not a number and anyway, her teacher could not know it was one million because there were only ever just a few creatures coming out of the holes. But now that she knew, Tebogo saw small holes everywhere and started to count them, writing down the numbers in her school notebook. On her forays with her mother into the reserve, she forgot to wish up a lion, absorbed by the holes and cones on the ground. There were always cones and holes but her creatures were not always bold enough to be wandering about. She thought that the most cones were in the bare patches where the grass was least. And where the cattle or goats had been, there were always plenty of holes and cones. Tebogo decided that there must indeed be many of these termites, many, many; too many to count and they must eat a lot of grass, more even than the cattle. Perhaps even one million.

On the days when she made it to her mopane perch her termites were not always there. Not even when she jumped down and brushed her hand across a sand cone. But when they were there she would count the grass stems as the little creatures shuttled along.

Tebogo never tired of her industrious animals. It felt good to watch them, for small and ugly as they were, somehow she knew that this gathering of the dead stems was important. It was part of how the world worked. Just to be still and watch them go about their business gave her a sense of peace that she knew all about without being able to explain. She said nothing to her friends.

It was not long before her rotund aunt returned the visit, waddling along to the boma on fat legs covered by a loud skirt. A shallow basket was perched on her head. All the children were called to greet such an important relative, the little ones still eager to see what treats might appear.

"Here" said her aunt as she took the basket down from her head. "I caught them yesterday after the big rain."

Winnowed to remove the wings and lightly fried, Tebogo had eaten these termites before and knew they were a good meal. The old folk in the village were especially fond of them. But these crispy things, scratching together as her aunt shook the basket, looked very different to her little creatures. These were the really big ones that came from the tall castles. She had been with her mother to the floodplain to find them. Her father said that they were bad things that had to be eaten; else they would eat the poles and the thatch of the houses in the village. Tebogo wondered if the fat termites in the basket that her aunt laid in front of the fire to be picked at by hungry fingers had eaten part of her aunt's house.

Tebogo said nothing about her termites. It was easier that way. Her aunt would only find something bad to say.

"Have you a boyfriend Tebogo?" her aunt said in a hurry between crunches of termite abdomen.

"No mma," Tebogo replied lowering her eyes to stare at the sand.

"Why not, why not, you are ready and the boys are many. What are you doing with your time young lady?"

"I am-"

"Oh she is never here," Tebogos mother said turning away from the youngsters. "Always out of the yard she is. She must be seeing boys. She is just shy about it."

"There, you see, no need to be shy. There are many good boys for you Tebogo. Your friend Sethunye has a boyfriend with a car. You will too if you go and find one. But I think you have one already. Yes, you should go and find one Tebogo."

"Yes mma."

Tebogo decided that it would be best to go along with what her aunt wanted. It was not wise to upset your relatives, especially an aunt who had eaten too many termites.

"Here." Her aunt said thrusting the basket under Tebogo's chin. 'Eat young lady, eat. They will make your breasts grow to the size of melons."

Tebogo ate a few trying not to imagine watermelons under her school dress. "Can I go mma?" She said to the sand in a whisper.

"Yes, off you go." Her aunt bellowed without reference to her mother. "Make sure you go and find those boys."

"Thank you mma," Tebogo said as she glanced at her mother who nodded but did not smile.

In a moment she was opening the gate as though she was eager to find the other teenagers in the village. Once out of sight she doubled back and ran to her perch, jumped into the cleft of the mopane tree, her breath fast and short.

Her eyes darted around for something to focus her thoughts, to slow them down. It was a blur. The clouds were in a race to reach the horizon. The trees waved on a stiff breeze. There were no birds. No mongoose. Nothing.

The only noise was the wind chasing past the leaves. The world was too big, too fast. Tebogo looked down at the sand.

She forced her gaze onto the nearest hole, a black dot on the crust of yellow sand. Nothing. The next hole was the same. She checked the dozen tiny piles of loose sand within easy view. Nothing. She settled on the last two holes, close together in the shade of a boscia bush, her chest still gasping and snatching at the wind. Nothing. She leant forward and turned her head to take her eyes back over the sand.

They must be here. They must be.

Still nothing. She flicked a hand at something rough that brushed her leg. And again she brushed her leg. She looked back.

Two huge eyes on an oval face were looking at her, a sinuous ribbon with a green tinge propelling the wide eyes up the trunk of the moapne and across her ankle. Tebogo screamed.

She kicked her leg and overbalanced, falling heavily onto the sand, sending the boomslang snake in the opposite direction. Scrambling to her feet she crushed some termite cones and caved in three holes in her panic to get away. She did not stop until she reached the boma.

"Back so soon young lady," her aunt said. "It is not good to be fussy..."

The words trailed away as Tebogo ran inside the hut.

Tebogo woke to the sound of her mother talking and laughing. Her aunt must have left. It was the neighbor talking to her mother, the disorganized but jovial woman who was round enough to have many termites inside her. Mma Raseroka always had something to say that was accompanied by a smile and a chuckle. Talking took up all her time so it was no wonder she couldn't keep up with her goats. Unlike those from her aunt, these words were soothing. They never left a bad impression, perhaps because amongst all the talking she was also happy to listen. Tebogo walked out into the afternoon sun.

"Dumela mma," Tebogo said in polite greeting to the rotund woman. "Your goats mma, they are far?"

"Em, Puso has taken them to the lands." The woman smiled. "You will not have to chase them for a while."

Tebogo glanced across at the pumpkin plants that were about to fruit.

"Our people have always had goats, Tebogo. They eat the plants that the cattle do not like and are easy to keep."

Easy for you, Tebogo thought, remembering the hours protecting the spinach.

"They do not hurt the plants even if plants do look different when goats are around. They are very clever and get to parts of the plant that other creatures cannot. But the plants always recover when the rains come."

Tebogo still thought the plants in the game reserve looked happier. Even the spiny dichrostachys bushes had more leaves. She had seen the impala eat some of the same plants as the goats but they never seemed to stay at the same plant for that long. Perhaps the plants hurt because they did not know goats. They knew impala, giraffe, kudu and eland, but goats were new to them. Just like you were sometimes wary of strangers from far away, so goats might scare plants.

"Tebogo." the woman said her name as though she had whispered it in her ear.

"Yes mma"

"Tebogo, you know my nephew Masego?"

Tebogo looked up to see her mother was beaming as she shuffled a little closer to her neighbor. This was not a good sign, as a smiling mother usually meant some mischief. "Em mma," she said.

"Well, he will be here today. You know that he has started to work in his fathers shop." This explained the smile.

Tebogo smiled too. Masego was thoughtful and kind. He was still a boy, not really a man, yet he was polite and always immaculate in his dress but not flashy like most boys. She liked Masego.

Masego liked Tebogo. As time passed he began to visit his aunt often, so often that everyone was smiling. The pretence was soon dropped and Masego came by just to see Tebogo, politely greeting his aunt as he passed. Tebogo's mother smiled.

Masego was very kind and attentive to her. The perch on the mopane tree was forgotten. Tebogo even lost the image of the bulbous eyes on the oval face.

Masego had brought Tebogo many gifts including a mobile phone for her birthday. It only worked in the village if you climbed a tree on top of the sand ridge but she would climb it regularly to collect messages. Any free time she spent in town. She went to the mall with her old school friends, most of them now working in stores or for safari companies. Masego would often buy her lunch. Chicken was everyone's favourite. He was doing well in the shop and Tebogo lost count of the times her mother mentioned marriage. She would promise at least three goats for the wedding feast.

Tebogo married Masego and had two children of her own. She would take them to visit their grandmother and sometimes, when the spinach plants were greenest, the wind was just so and the fluffy clouds made her look up at them, she thought of her little creatures. She remembered what it was like to be with them. To be on her favourite perch on the mopane tree, to watch and count them as they carried tiny grass stems. Warmth came over her, a glow from somewhere unknown. She heard inside the chant that used to keep the lions away.

She wondered if she should go and take... *No. Not there.*

Chapter 12

Coda

The awkward news is that we do not touch the earth, we stomp

I sleep on a bed. It is constructed to support my body in comfort with a cushioned mattress on top a base that raises the height to make it easy for me to get in and out. I live in a bungalow built on a slope in a way that creates a great view of a rainforest gully from our back garden and puts the floorboards of my bedroom several feet off the ground. I can easily crawl under the house. When I am in bed my body is six feet above the earth.

The pathway from my front door to the garage is paved, my car is upholstered, the path from the car park at the train station to the platform is also paved and the train runs on steel tracks. I can sleep, wake, wash, eat breakfast and get to work in the city without having physically touched the earth. And so I don't. Modern lifestyles rarely include time when our bodies are in contact with nature.

This is not healthy for us. Our spiritual wellbeing is compromised by our detachment from nature, as is our ability to understand the complexities of our effects on the environment. Yet we are not built to understand the full workings of nature. Our basic biology is fast in the scheme of things. The thought experiment in *Taking your time* probably took some following because our normal frame of reference is so short.

I never met my great-grandfather and I have only vague recollections of my grandfather who died when I was six. Short-term memory is just as sloppy; it is a struggle to remember what I had for dinner two nights ago, let alone Monday last week. Memory is useful to us but our instincts and brain function operate best in the present. We are even encouraged to find enlightenment in the here and now. So when we find out that more-making happens over decades, succession and population processes over centuries, climate and geomorphology over millennia, evolutionary process longer still; it is little wonder we construct our world without a full appreciation of these things. The environment is not on our wavelength, and it is possible that we do not have the capacity to be on its.

Our preponderance is also for the world to be linear. Flick the light switch and there is light, because we like it when an action leads to a result, preferably an immediate one. This requirement dominates our decisions on most things, including our use of natural resources. We imagine that if we reduce carbon emissions, then we will avert dangerous climate change. There is no place for contingent effects, feedback loops, synergies, confounders or, heaven forbid, chance. Nature, by contrast, is interconnected. It is a web, with linearity rare everywhere except in our technologies. Again, what is and what we prefer are not aligned.

In *Ants* our terminally ill scientist is lost. He has trained hard to deliver linearity and explanation through cause and effect, yet he knows that as the world changes so some things stay constant in spite. In an odd way this reassures him. The fact that ants can forage on anything, including his lunch, and can go about their business without fuss means that they must persist beyond him. The human approach is to modify the environment rather than adapt ourselves to it. Even though we may believe otherwise, our actions are out of sync with the vibration of nature. At times it seems that our arrogance and overreliance on linear thinking, will be the end of us. For, after all, what if it is not emissions?

Tebogo was young and free and in awe of her *Little Creatures*. She could see nature and had an affinity with it despite the fact that goats were a conundrum. But, as it does for most people, her connection would pass, it would not be her life. That would play out somewhere else away from the silliness of tiny insects. A scare was all it took. She lost her touch on the earth. Sadly, in our current electronic world, many children do not have a connection to loose.

Since the middle Ages humans have increased in number, exponentially since the industrial revolution, and grown their material need at rates that seem unable to slow. Our collective touch on the earth is more like a slap, or perhaps a stomp. Human *beings* are not. We have forgotten how to be, mainly because we do not touch the earth.

There will be a shift in the way we do things. It may begin as awareness for our personal touch on the earth, the spiritual benefit of being in nature, then lead to knowledge of our collective touch and, soon enough, to conscious choices that are life affirming.

Part Three

Collective

Chapter 13

Overview

Collective

Rules

There is an ant crawling over the keyboard. Chances are that there is one within a few feet of you as you read this book. There will be a spider too, but do not fear, they are both there all the time.

Ants are ancient creatures, first appearing sometime in the Cretaceous period around 130 million years ago, long before placental mammals, and since then diverging into thousands of types with at least 12,000 species alive today. They live in social groups that range in size from a few dozen predatory individuals to highly organised multitudes. An ant colony can occupy a large territory with millions of mostly sterile workers, all scurrying relentlessly for the common evolutionary good. Ants live on almost every landmass on Earth. They thrive in most habitats and may make up a quarter of the animal biomass.

An entry in Wikipedia says *'[Ant] success has been attributed to their social organisation and their ability to modify habitats, tap resources, and defend themselves.'* This description is eerily familiar.

Ant colonies function well due to a simple division of tasks built up around a small number of decision rules. While we might like to imagine that morning meetings, support groups and production schedules go on in special chambers underground, individual ants have some pretty basic programming. Yet when many of them combine, there is an exponential increase in outcome. A lot of individuals making responses to a small number of decisions create the appearance of mass organisation.

Humans use similar rule-based logic to ants. Go when the traffic light is green, stop when it is red. It doesn't matter what the person in the yellow van with roof racks is thinking, or even where they're going. They need only make a simple driving decision based on the colour of a light. If they obey this simple rule, then the traffic will flow without incident. Viewed from above the movements of the vehicles will appear highly organized.

Humans have taken the decision rule idea to great lengths. We have rules for everything. The reason we can have them is that we have the ability to learn and the systems of communication to store and transfer the rule sets. Most of the animals our size lack this complexity and would not do very well at the sorts of densities we exist in. Can you imagine six billion dogs, some of them living in cities of 15 million? It takes pretty sophisticated communication to hold it all together but more than that, it takes rules.

Most animals rules are actually behavioural triggers – if this happens, then do that. But as we discussed earlier, humans are sentient beings with the ability to consciously change our surroundings. So for us, as well as a set of behavioural triggers, we have rules that are evolved constructs. That is, we made them up.

Rules are needed wherever there is potential for conflict. In sparse populations where interactions are rare and resources are plentiful there is little need for regulation as conflicts don't regularly or normally occur. Situations where I don't run into you much and you don't have anything I want anyway can be handled with a few simple rules. We agree that this is our turf, and that is yours. We can make this agreement with a few simple displays of strength and intent. Then, as long as we stay off each other's patch, all will be well. We can both spray urine on the boundary, to ensure we both know where it is. And if you want to come across onto my side you need to ask first.

The invention of agriculture around 10,000 years ago really set humanity underway in terms of growing population and with it, growing rules. We learned to work less as individuals and more as a collective. All of a sudden the time we needed for activities to keep us alive was reduced; we had time to sit around and make stuff up. And as we know, life is amazingly inventive.

Anthropologists point to "specialization" occurring at this time. That is, if you guys over there look after supplying food for us, then we'll make your tools, that lot will keep you clothed, he will make up stories to make you laugh, she will work out what keeps you healthy and those people will keep God off your back. As the various human societies developed, so organisation presented in various ways, think of say, the Incas, the Egyptians, the Picts, the Pacific Islanders.

As the numbers of people increased and farming allowed them to live close together, the need for rules grew. Social organization emerged to manage people living in close proximity.

Two broad rule sets evolved. One came from the traders, business leaders and lawmakers. These rules were responsible for solving conflicts related to survival such as; he stole my money, he is selling too cheaply, she's on my land, he killed my son. Another rule set came from the need for emotional and spiritual well being - he stole my wife, she looked at me funny and I got sick, why am I so unhappy, they killed my son, and so on.

This latter set of rules harnessed faith and belief and became a critical element of social organisation. It was necessary because the trading rules were not sufficient to cover the things that were emerging to fill 'free time'.

Rules evolved and persisted because they ultimately made existence easier. Life was more comfortable, more resource rich and safer. Clearly no Inca or Easter Islander could have dreamt of a bed with springs, air-conditioning units or aspirin, but relative to the alternative at the time, collective effort made it easier for humans to stay alive and make more.

And the rules helped.

It did not make sense to be loosing members of your tribe because of 'small' issues – just because he shook his spear at you doesn't mean you can kill him. We could make him take it back or punish him in a way that makes you feel happy again and yet still allows us to keep him alive and useful to society. He would be especially useful with his spear when the nearby tribe attempts to steal the maize crop.

The rules have continued to evolve. Nowadays, even in times of conflict over resources we are much more reluctant to go to war than we have been in the past. Medieval collectives would fight for honour or to defend the king. In the first world war many Australians signed up to defend the British Empire and came back heroes; but by the time of the Vietnam war in the 1970's there was public outcry at young men dying for another country. The rules are now complex enough for even scarce resources to be traded, for the most part without bloodshed. Scarce goods simply command a higher price or come under a trade agreement, treaty or bilateral arrangement.

There have been many large tomes devoted to our systems of governance. Just like evolution creates diversity with time, so the millennia since humans first began living together in large groups have seen our rule sets become hugely complex. We have seen many that work, but just as natural selection weeds out the less suitable types, so some rules have expired. We no longer punish women by burning at the stake for breaking the rule that only witches take a walk in the countryside.

Among the copious rules that have evolved to support human organization in modern societies, there is one set of rules that are curiously absent. That is, rules about the environment. Anyone still under the age of forty may not realize that it was not until the 1970's that legislation to protect the environment began to appear on the statues of western countries. So, while we've been manipulating the environment for 10,000 years, it is only in the last 40 that we have thought to provide rules about how to do it.

In the popular clock analogy, if we have been using agriculture since midnight, it took us until 23:54, almost a full 24 hours, to make the collective responsible. Even if you start the clock at the industrial revolution, around 1850, it still took more than 120 years or until 18:00 hours. There were some 'early' concessions to parks and reserves in the past hundred or so years but even today, effective legislation in many countries is still absent or incomplete. For example, it is only in the last five years that Australia has taken a firm stance on land clearing.

It is easy to argue that we didn't need rules for most of our past. There just seemed to be endless resources; the environment was not constrained. When the early European explorers reached Africa they must have believed natures bounty to be endless.[36] It would have seemed crazy to think of restraint and impossible for the resource base to be affected by our actions. Lance Armstrong, an old farmer from in the central west of NSW, Australia, talks about what happened in the area in the 1930's, some 100 years after the first Europeans arrived.

> *"We never thought it would end. There were bloody trees everywhere. You couldn't see the sky. We just used to clear a little bit more each year and all of a sudden there wasn't much left".*

Yet there have always been some who were concerned. Dick Bland, a beekeeper from western NSW was passionate about over-clearing of the inland. Not just because clearing removes good honey country but because of the decline in what we now call ecosystem services. On a visit to a property east of Bourke in 1997 we stood on a bridge after driving past thousands of hectares of bushland poisoned or cleared to run sheep or to grow cotton.

> *"We used to drink water from that creek when we were chasing honey in 1954. Now it looks like pea and ham soup."*

As we have discussed, humans are highly effective in our ability to modify the environment to suit ourselves. This has lead to the downfall of other animals and plants. The British Isles only has a fraction of pre-industrialization biodiversity. Cyprus, once covered in magnificent forest is now mostly

---

[36] It is sobering to note that a flight on a Kenya Airways Boeing 707 into Entebbe, Uganda in January 1989 happened only 37 years after the airport was opened and just 130 years after the first Europeans 'discovered' the Buganda Kingdom.

cracked, dry and barren. Some islands, populated for thousands of years, are now incapable of supporting the resident human population without importing resources.

Yet in some places micro-management of the environment has led huge and efficient forms of primary production, sufficient even for the global collective. In the Netherlands a highly mechanized agricultural sector provides large surpluses with exports of food products and flowers earning $55 billion annually. Production includes a quarter of all world tomatoes and one-third of the world's exports of peppers and cucumbers. Given this and close to 5 million cattle and 13 million pigs, there is very little space left for anything else.

It seems that whatever the first impressions of plenty, the effect of the collective has been to see resources used and at a cost to other organisms. There could be an accusation that the absence of strong rules around resource use by our urban world has contributed to the current global extinction trend. And there may be a case to answer.

Should the case go to court the prosecution would no doubt call for evidence from elders of human societies that do have environmental rules. They could be called from Aboriginal communities on all continents. They would say that indigenous people knew the limits of the resource and put their own cultural rules in place to make sure that the limits were not exceeded.

Anthropologists and archaeologists would also be called as witnesses. They would say that Australian Aboriginal communities, for example, spent about 3 hours per day looking after the need for food and shelter. That leaves about 7 to 11 hours per day to do... other stuff. In the absence of a need or desire for technological innovation indigenous societies used this capacity to develop socially. Even a cursory study of Aboriginal culture reveals huge complexity in the rule set.

The anthropologists would argue that the first Australians were rare in their ability to manage the human traits of greed and avarice, because the rules in Aboriginal society were often strongly linked to the environment. Like the American Indians, the concept of a totem, relating an animal or plant to groups and individuals, carried with it rules and responsibilities around that species. These rules were as fundamental as what you could eat and whom you could breed with. The rules ensured the prosperity of that species and, because everything is connected, the whole system. We know, the specialists would conclude, that Australian Aboriginal society prospered within the productive capacity of the land without causing the extinction of one third of mammal species and thousands of plants that has occurred since European settlement.

The defense would bring forward its own evidence. All very well to have a rule set for the environment when there are only ever a few hundred thousand people, 1.25 million at most across a huge continent. If it did not work then, when would it? Plus these people lived in small bands. Yes, they had nations and complex cultural identity but they did not live an urban life.

Then, of course, there are the extinctions of mega-herbivores and the alteration of the landscape by the fire stick. Perhaps this was the reason the cultural rules on the environment came about. These early peoples did just what we are doing today in our legislation; they fixed a problem they created.

The debate would go on. The jury would find the evidence hard to digest for there would be logic on both sides.

In his summation the judge might conclude that it is open for debate as to whether resource use constraint was a deliberate choice by the aboriginal people. It may have been a consequence of a set of rules that emerged aligned to the capacity of the environment to support the human population or the rules may have emerged because they noticed resources were in decline. In the end they did not have to deal with food, water and shelter for 21 million people, so their rule sets were very different. The case for the absence of strong rules contributing to the current global extinction is one for your conscience.

The group effect

Another relevant aspect of human society in this debate is that of group theory or, how we behave when there are lots of us together. I once met a researcher who was comparing the behaviour of seagulls when thrown hot chips to that of humans when free products or money were handed out. As you may visualize, we were not that different!

The behaviour of individuals in groups and groups themselves in a contemporary context is fascinating. Let's look at a few examples:

Buffet Theory: Most individuals will overeat and will jostle for prime position, often feeling disappointed or inferior when they don't get it.

Lift Theory: Forget the mob, it's all about me. Most people think that the lift is just for them and hence, will run into people coming in or out.

Baggage Claim Theory: In the absence of rules, individual self interest in claiming baggage will result in group dysfunction or other individual discomfort as everyone crowds around the carousel. If we could all just take two steps back!

The reason for talking about this is that for societies to work we need to overcome certain 'default programs'. Managing the "fight or flight" instinct in large groups is a real concern for organizers of major events or any collection of people. Managing the hoarding instinct is a big challenge for governments that want everyone to have a piece of the pie. It is one of the big challenges to overcome for achieving the desired outcome of shifting the global economy out of recession. When people are fearful, they hoard. They do not spend. Unfortunately to sustain economic growth people will have to keep spending.

Getting over your own fundamental self-interest for the good of the whole group is a challenge. It is essentially what green groups are asking us to do. It would appear that we struggle to do this now, when times are good. How are we expected to do this should resources become constrained? In reality they already are constrained because not all of the planet can live like us.

In the previous section we proposed that becoming more aware of your connectedness to the environment as it really is will enlighten you and give you a better quality of life. The extension of this as a collective is that if everybody did it there would be greater quality of life for all. If we become more enlightened and connect more closely then we will collectively make decisions for our environment as well as ourselves. Individual decisions will lead to an increase in our collective quality of life. It is a positive feedback loop.

Up to now the conservation movement has been focused on telling us how much we have damaged the environment. Actually, the environment doesn't care. It has just changed in response to our impact and will continue the same processes long after we have gone and with whatever compliment of life that is left. It really only matters to us. And so far, as long as our production systems have supported our own needs we have not worried. But we are beginning to see that the edges are fraying, that our land might not be as productive as it once was. We are not sure that there are resources capable of supporting 9 billion human engines.

In a sense, the environment movement has made a rod for its own back. The issue has been defined as a problem **with** the environment. A problem that is external to us. While ever we continue to relate to the world as if we are not part of it – be in the world but not of it - we will struggle. The quality of life for future human generations depends on how much connection we can engender now.

A future we choose will not occur by doing more of the same. Clinging to the past will not help. Our conservation legislation is totally inadequate to take us forward with a growing population. Biodiversity reserves and individual species recovery is not going to inform how we are going to produce food and resources for 9 billion and counting, all wanting more than US$10 per day. The only way to do this is as a collective but the only way to the right answer is to have individual connection.

We will need a biodiverse landscape with a smaller percentage of the land area producing a greater amount of our food. This will require major structural changes and it cannot be dependent just on the dollar. It's about making strategic decisions for 'what' and 'where' based on land function, energy, water and nutrient balance. In this case we would value biodiversity not just to have more cuddly critters, but because we acknowledge that biodiverse systems have the maximum capability to keep systems running hot.

Although some individuals will run their farms on this basis, we cannot make these decisions just as individuals. We must achieve a collective that says 'we need to change', or we will be in deep trouble with war, strife, pestilence, disease – all the mechanisms you would expect of density-dependence when resources become limited.

A look into the future of agricultural production does not require a crystal ball; we need only look at what has occurred so far. We have massively altered landscape over huge areas. Soil fertility is declining. Fertilisers are needed to maintain production levels and the fertiliser mines are running out. Two thirds of the worlds bore fields are running out of water. All the time population is increasing. Do we do more of the same to meet increased need? If we do, then more using and altering of the landscape will deliver continued land degradation and declining fertility. And eventually the resource will be become limiting.

If we were connected we would fit the commerce and the politics to the environment – not the other way around. At the moment our choices in land use are driven by commerce, then politics and then environment. As a group we would be capable of making a decision that would reverse this status quo. If we connect with the environment, our decisions around resource use will be closer to truth. For example, we will value biodiversity, not because we feel an emotional attachment to a particular creature, but because we recognize the importance to us of a diverse landscape.

It is a choice. Ultimately an individual choice, and if we made it the system would change.

There is another aspect to group theory worthy of discussion, that is agreement or consensus. It appears that once a critical number of people do or believe a certain thing, it becomes known and acceptable to the entire group. The fads and crazes that we see grip society are a good example of this. Initially the idea or product is individual. As the number of individuals seen to accept this idea or product increases there is an exponential rise in consensus. Awareness also grows on the basis of consensus and ultimately, at least in a democratic society, has the power to influence the present.

An illustration of consensus relating to the environment occurred recently in Australia, we can call it Kevin's Dilemma. Before the 2007 Federal election in Australia the news was filled with stories of climate change. Al Gore's film "An Inconvenient Truth" had captured the imagination and, according to popular polls, Environment, for the first time in Australia's history, rated as the most important issue worrying people.

The signing of the Kyoto agreement, whether right or wrong, became the focus for many voters and, along with an apology to Aboriginal people for past atrocities, became a major point of difference between the incumbent Liberal party and the Opposition, the Labour party led by Kevin Rudd. The polls were right. Kevin Rudd had the greatest alignment and won in a landslide over the party that had denied climate change and refused to say sorry.

After just over a year in office, the Australia government has recently declared its intention around a Carbon Emissions Trading Scheme and has set a 5% target in reductions. It would be interesting to imagine what this number would have been had the policy been set before the election. The big issue now is the "economic crisis" and with few people banging on about the environment the Prime Minister was much more ready to listen to the conservative economic rationalists who lobbied for 5% than the greenies who called for 60%. Had there been thousands of people on the lawns of Parliament House the result would probably be different.

I suppose the thing to take away from all of this is that you are connected to everyone and everything else and part of a collective which has great power. Because of the way we are organized you can imagine that everything you do and say is important.

In a world of increasing population and increasing scarcity of resources perhaps one of the most critical concepts to grasp is also the most simple: You are the collective. You are the problem. You can also be the solution.

Chapter 14

Essay

Journey to a Farm

What our landscapes really look like.

Australia is a big, isolated place that has been drifting across the ocean for 45 million years. In a way this, and the two significant arrivals of humans (50,000 and 200 years ago) is all you need to know to explain the characteristics of this island continent. Age, size and isolation define what it looks like and how its biology functions, to how it will fare in the future.

The land area of Australia is more than 7.7 million km^2 making it the 6th largest country in the world, big enough to be a continent and famously an island. The landmass has been separate from any other since the middle of the Eocene epoch, 45 million years. In the Eocene a moderate, cooling climate enabled archaic mammals to flourish, the first grasses appeared, and vast quantities of seafloor algae decayed. This decomposition on the seafloor pulled massive amounts of carbon dioxide from the atmosphere lowering concentrations from 3800 ppm down to 650 ppm. Around this time the Antarctic ice cap reformed, an event that would trigger the 'Icehouse Earth' climate that followed to the present day. Whilst all this was happening the landmass we call Australia began to drift north as the plate on which it sits moved to bump into Asia.

In the Oligocene epoch from 40 to 23 million years ago, Australia continued to drift north and east as the climate cooled. Evolution took on a spurt with rapid diversification of organisms, especially mammals and flowering plants. Then, in the Miocene and Pliocene epochs that followed most of the recognizable animal and plant families appeared as widespread forests drew more CO_2 from the atmosphere to lower concentrations from 650 to 100 ppm. Then, in the 2.5 million years of the Pleistocene epoch large mammals flourished and became extinct[37]. Humans appeared at this time but did not reach Australia until around 50,000 years ago (there is debate as to exactly when) in the Holocene, the current epoch.

In this time of drift there would have been many changes to the climate along a generally cooling and drying regime. The Australian Bureau of Meteorology talks of the 'Federation Drought 1895-1902' and 'The 1914-15 Drought' as significant events. True enough at the time for the farmers and graziers trying to produce through them, but insignificant in terms of the pre-instrumentation records. The evidence from ice-cores, pollen records and the like suggest climate was much drier, colder and windier than these 'droughts' during past interglacial maxima. There were also times when there was more moisture around than today. For the most part large size, position on the globe and the global climatic regime kept the drifting mass relative dry and windy as it went on its journey north.

Before it started to move, the landmass of Australia was part of the much larger Pangaea supercontinent that formed 250 million years ago. It was the breakup and drift across the globe of this huge area produced the current continental configuration. The weather and a layer of biology had worked the land for a long time before the 45 million years of drifting. The period of drift was, in itself, long enough for mountains to form and erode, rivers to cut and deposit, soils to develop and erode away. This history means that the landscape in Australia is very old. It has been a very long time since the land was inundated, uplifted or refreshed by volcanic activity.

A flight from Sydney to Auckland in New Zealand tells this better than words. You leave the flat basin of Sydney with

---

37 A famous example is the genus *Diprotodon*, at least two species of marsupial herbivore about the size of a hippopotamus that lived in forests, woodlands and grassland of Australia for just over 1 million years only to go extinct about 40,000 years ago.

the gentle hill slopes of the Blue Mountains in the distance, to cruise over the ocean. Behind you is smooth, flattened, land as though some giant steamroller had been back and across it a thousand times, unable to get all the lumpy bits out. Then after a about four hours in the air, the aircraft descends through the clouds to reveal jagged peaks and gullies that look like they have just been thrust up from the ocean. In geological terms the islands of New Zealand have been around for about 500 million years, rather less than Australia that boasts some of the oldest rocks in the world at 3.5 billion years. The real difference is that last 10 million years in New Zealand has seen tectonic activity and uplift that makes the whole place look new.

No such recent geomorphic renewal has occurred in Australia so the landscape is one of shallow erosion, gorges, rolling hills and flat deposition plains. Throughout this old scene the soils are heavily weathered, a consequence of thousands of wet/dry and cold/warm cycles eking out from the parent material the mineral nutrients so essential for the plants. Weathered soils are generally nutrient poor, often low in organic material and have a weak structure. In short, they are easily disturbed, especially by novelty. So when new disturbances arrived in Australia, first in the form of more frequent fire and predation of large animals (some time after 50,000 years ago) and then mechanized habitat alteration (at some time after 1788), the soils and the vegetation they support, had limited resilience. There was only a weak buffer in the form of soil carbon, thanks to the dry climate that had persisted for epochs. So even though the compliment of plants and animals was well adjusted to the dry, the soils were not ready to have their vegetation removed. The human impact notwithstanding, the picture of Australia before humans arrived is a big, isolated place with a landscape shaped by great age and a mostly dry climate, with warm, windy conditions.

Biodiversity is a term to describe the variety of life, the multitude of forms, life cycles and ways of more-making that are the current products of evolution. Australia has a disproportionately large share of unique ways of more-making. Time and size are enough to create diversity and isolation ensures that a large proportion of it is unique. Nowhere else have marsupial mammals radiated into almost every niche.

Elsewhere the placental mammals have the upper hand. Nowhere else have ants found the dryness so appealing that they diversified into a myriad of forms or the vegetation had such a long and consistent time of dry, nutrient poor times to have come up with types of plants that not only handle these conditions but the frequent fires that go with them. It is a picture of serenity.

There has been time enough for the natural processes we imagined in our earlier thought experiment to have etched life onto the landscape. Time enough for the trees to grow, flourish, die out, and evolve forms well suited to the dry and the frequent fire. Time enough for ants to have co-evolved with specific plant species so that neither can survive on their own. The conditions may seem harsh to us, but not so difficult that nature couldn't find a way.

So what happened? We are told that this unique biodiversity of Australia is being lost at rate faster than anywhere else in the world. This is in spite of a perception that green consciousness is strong among Australians. We are told that continent is drying, that climate change is upon us and will, among other things, further accelerate biodiversity loss.

We are not always told that human impacts since 1788 have radically altered the landscape in order to provide resources. There are currently around 21 million people in Australia most of whom live in the cities and around the coastline, leaving roughly 1.5 million rural dwellers to look after 28 million cattle, 79 million sheep (down from 120 million in 1997), 97 million chickens and 482,300 km^2 of cropland.[38]

The Australian economy generates an annual GDP of over $900 billion ($43,000 per capita); activity that grew by 31% from 1998 to 2006. In economic terms average household consumption in 2006 was $26,100, up from $10,400 in 1961, and has risen sharply since 1996. Up until the global financial crisis of 2008, by most measures of economic success at least, Australia was doing very well.

---

[38] Relatively few people manage agriculture in Australia – there were 150,000 registered businesses with agricultural activity in 2007 and given the universal rule of thumb that 80% of activity comes from 20% of participants (Pareto Principle), we can expect that most of the agricultural production rests on the talent of 30,000 people.

This economic activity requires resources and power, so with economic growth comes increased resource use and associated waste. We have already seen that domestic energy use doubles every thirty years and solid waste from economic activity amounts to 1,600kg per person per year. Feeding the people in a dry land means that agricultural activity needs to use around 12,000 gigalitres of water each year for irrigation, over two thirds of total human water use and enough to fill Sydney harbour 24 times. Even though the continent is huge, 58% of the land area is used to feed people and animals. The estimates are that in 1788 a third of Australia had forest cover; in 2003 cover stood at 21%, a reduction of over 900,000 km^2 or slightly more than the land area of NSW.[39]

Place these two stories together - a long period of evolution in isolation and the success of modern humans - and the reasons for rapid biodiversity loss in Australia are easy to see. A system subject to slow change over eons was hit, first with a fire stick and then a tractor and millions of sheep. Whenever such novel disturbance is introduced to a unique biota there will be losses. It is inevitable.

Biodiversity loss will happen when one species appropriates an increasing share of resources. It has to be so because there are fewer resources left to go around. And yet we seem to have trouble with this reality.

Travel west

Most Australians live in urban centres where the resources they need for food, shelter and comfort are found in their local supermarkets or delivered direct to their homes by utility companies. These people go to work in exchange for cash that they use to pay for their home and the resources they use. They have little need to know, or to understand where the resources come from or the magnitude of human impact on the

---

[39] The rider on this agricultural impact is that arable land covers 7% of the land area (compared to the global value of 11.3%) and produces 70% of the daily per capita calorie supply for humans, plus grain for animal feed.

landscape necessary to maintain supply. Indeed, most urban dwellers do not visit the countryside other than to travel through en route to holiday destinations on the coast. Some sit on flights that cruise over the landscape at 30,000 feet, headphones on, gazing out of the window just before touchdown.

When urban dwellers are told that humans appropriate nearly half of net primary production from plants, it is just a statement. Primary production happens outside of our perception – in the countryside somewhere. When all comfort and entertainment is available in the city, why go anywhere else.

It would be good to take a look at this countryside. One way would be to encourage people to take a drive out west. Lets say a journey from the Sydney Opera House to Menindee, a hamlet in the dry country of the real outback, 1,100 km distant. The journey would take a couple of days and go something like this.

In the first hour of the drive, two if it's the afternoon peak, west from Sydney Harbour you pass through the Sydney basin, home to around 4 million people. The business district and inner city streets look suitably crowded and then the freeway scoots past a sequence of suburbs with spacious plots interspersed with the occasional factory. After about 70 km you reach an escarpment that looks like it is blue thanks to waxy leaves of eucalyptus trees on heavily wooded slopes playing tricks with almost any light. It takes an hour drive across these Blue Mountains. Not high, they might pass as foothills in Colorado, they are dissected by numerous gorges and it took European settlers almost half a century to find a pass through them. There still are only a couple of roads across these mountains.

The villages along the route are a delight and boast more coffee shops and bakeries per kilometer than downtown. Refreshment stops for the many people who like to visit. Push on and soon the road takes you down the western side of the great divide and into mixed arable and grazing lands. Occasional fields of cotton, grains and legumes sit alongside grazing paddocks, all stitched together with threads of woody plants. On occasion, there is a remnant of natural vegetation hanging on thanks to rocky soil or outcrop.

A short detour just past Bathurst and you can slide past Cadia Gold Mine, a 1km wide hole in the ground on a site that covers close to 16 km^2 of the countryside with spoil and tailings dams. The mine is the second largest open cast pit in Australia after the huge super pit at Kalgoorilie and is expected to produce gold and copper and contribute to the local economy to 2013. It's a thirsty spot as metals are extracted by floatation after grinding crushed ore. Water supply is an issue here as there is a limited supply and demand from agriculture and local communities in Orange and Bathurst.

Keep going for another few hundred kilometers and slowly but definitely the vegetation shrinks. Trees become shrubs and grasses thin to leave the soil more exposed apparently happy to show off its red hues. Now the paddocks get bigger. This is sheep country. In NSW there are 27 million sheep, roughly a third of the total on the continent. One hundred years ago there were 200 million, an order of magnitude greater than the current human population.

Press on, it's a real journey this one, across the featureless plains peppered with sheep, emus, the occasional cow and crepuscular kangaroos. All the vehicles out here have serious ironwork to protect them from these none too bright herbivores. Now we have reached a dry place. Average annual rainfall in West Wyalong is 478 mm, just a third of the amount that falls on the Opera House. There are also only 78 rainfall days, half the Sydney number.

Stick at it and the road takes you to Hay, another country town on the Murrumbidgee River. It is still dry but there are arable fields out here on the floodplain fed by irrigation water. Take a right out of town and head towards Ivanhoe and eventually you'll see some signs to Menindee Lakes, hardly credible as, except for the glimpse of the river at Hay, there has been no surface water for hundreds of kilometers. It is also 38°C in the shade. In these parts for four months of the year the average maximum temperature is over 30°C. Take the left turn and travel on the narrow road to Menindee where the 980 good citizens expect 242 mm of rainfall per year on just 42 rainy days. This last section of the drive is a weird experience as the land is so marginal that you wonder whether it ever rains, then suddenly you come across vast freshwater lakes full of dead trees and surrounded by sand, saltbush and inhospitable red soils.

There are some 20 lakes around Menindee that fill when there is sufficient rain for a big flood in the Darling River. One of these is Lake Tandau, a shallow basin, 10 km across and 18 km long with an area of 17,000 ha. When floodwater reaches this lake it no longer spreads out but is channeled into dykes and pumped onto arable fields. Huge expanses of carefully prepared soil, grey like you would expect tilled soil to be, not red like the surrounding dunes that on winter days you can see on the horizon in all directions.

This is Tandou Farm, a multimillion-dollar agribusiness listed on the Australian Stock Exchange. In 2002 the farm grew 2% of the national cotton crop, some 50,000 bales, and also grows grapes, apples, apricots, durum wheat and malting barley thanks to the purchase of more than 33,000 mega litres of water licenses. Water for these licenses is dependent on floods that must first fill other lakes. The last flood event was in 2003, and the one before that in 2000. Shareholders pray for rain.

Tandou is a modern operation. Agronomists monitor the crop; the soils are tested and the water used carefully to maximize plant growth. Crop dusting airplanes sit in hangers on site ready to drop pesticides at the best time. There is even a small village for the workforce.

To many Tandou Farm is an emerald of arable production in an ocean of dusty rangeland. Others are less enamored. The following is an extract from a report by journalist James Woodford in the Sydney Morning Herald on 8th February 2003.

> *Until the 1950s, the Menindee system was one of the continent's most important wetlands. In wet years millions of waterbirds bred there, rivalling places such as Kakadu.*

> *The report reveals that more than 99 per cent of the lake beds have now been degraded by either too much or too little flooding. Levees, regulators, dam walls and pumps divert or control the Darling River water that once ran free.*

Successive governments, under pressure from irrigators, have managed the lakes in a way that Dr Kingsford says is the "antithesis of conservation".

Lake Tandou is a good example. Since irrigators took control of it in the 1970s, it has ceased to be a lake and 17,300 hectares of wetland have been destroyed.

"The lake bed is now used for irrigating crops, supplied by water from the Darling River," Dr Kingsford and his team write. Before the irrigators arrived there were 100,000 to 500,000 waterbirds, of up to 40 species.

Now, a few decades since the lake was handed to irrigators, fewer than 50 birds survive, eking out an existence in cotton irrigation channels. No environmental assessment was done into the loss of the wetland. The destruction nearly all happened in the past three decades, and continues, to the cost of the national park.

The excessive use of water by irrigators is also claimed by some to be threatening the water supply of Broken Hill. The city's main supply is the Menindee Lakes system, and it is facing the prospect of having water transported from South Australia.

The only way to restore ecological health to the Menindee Lakes is to allow the system to regularly flood and dry as it did for thousands of years.

It seems unrealistic to ask people to make this journey or any similar one away from another major city on the planet. It takes time, costs in fuel and wear and tear on vehicles and in the end there is not that much to see. The idea that by scooting by we can see what we have done to the landscape seems naïve. The view of the landscape from the car is not enough. Yet thanks to some amazing technology, not only can we make the same journey to Lake Tandou without leaving the suburbs, we can get to see the landscape as we go. It is possible to make this journey virtual.

Start by launching Google earth. Key in the 'Sydney Opera House' in the 'Fly to' box and the screen will show you flying around the globe to zoom in to the CBD of Sydney, the iconic building lapping at the waters of the harbour is at the top of the screen. Turn off all the various layers and zoom out to an eye altitude of around 10km (the counter is in the bottom right) and you will see the landscape. Features are still visible, the outline of the CBD, the Sydney Cricket ground in the south-east and the densely packed inner city suburbs. There is no primary production here beyond what is ornamental.

Now we can take a short 20km hop west by typing 'Parramatta, NSW' into the 'Fly to' box and again zoom out to an eye altitude of around 10km. The trip was across more suburbs and the scene we see is more of the same, save for the addition of some industrial areas and a slightly greener tinge thanks to the gardens in these more spacious suburbs. Another hop takes us west to 'Penrith, NSW', the edge of the Sydney basin and the western limit of the metropolitan area. Zoom out to 10km and there are a few gaps between the suburbs, fields are recognizable and maybe 10% of the landscape delivers more than space for buildings, gardens and roads. At this point the Nepean river skirts the foot of the Blue Mountains, the lower slopes are covered with trees.

The next hop of 35 km takes us to the tourist town of 'Katoomba, NSW' in the middle of the mountains. The landscape view is of a small town perched on an escarpment surrounded by forest. Indeed the virtual flight from Penrith gives us a view of similar bushland all along the route, with the distinct and uninterrupted pattern of drainage lines and ridges. The value in this undisturbed country is recognized in its status as world heritage area. The extent of this forested wilderness was easy to miss on the road trip as most of the settlement is in a ribbon along the highway, but north and south of the road and there are vast expanses of nature behaving as it has done for millennia. The mountains are part of the great divide, a spine of high country that separates the east coast of Australia from the vast Murray-Darling Basin. The next hop takes us 42 km across to the western side of the range and the small town of 'Oberon, NSW'.

As we descend from the mountains the landscape changes to cleared paddocks, most of them for sheep and cattle grazing with the occasional brown or sand coloured arable field. There are odd-looking dark lines where trees mark paddock boundaries. Many of these trees will not be native but dense foliage species planted for windbreaks. The forested areas in the north-east of the 10 km aerial view are plantations, mostly of pine and it takes a closer look to see the occasional remnant of the original forest. This is what countryside looks like wherever there is rainfall and soil good enough to grow plants for animals. The vegetation is modified into packets that humans can manage.

When rainfall is more reliable and the soils more fertile, then the packets get smaller, more rectangular and we grow crops. The next hop of 41 km to 'Bathurst, NSW' shows this very nicely. East of this sizable population centre along the river floodplain there are fields with their crops in rows, to the west the rolling hills return with bigger pastures. Humans are expert at knowing where to get the best return from the primary production. If an area can grow a crop, then chances are we have tried to grow one.

Although appropriation of primary production is the activity with the biggest effect on the landscape, there are also other resources that we access. Type in 'Cadia, NSW' and we make another short hop of 54 km to the hole in the ground that is Cadia gold mine. The 10 km view of a turquoise blotch on the grazing landscape, and to see it all it is necessary to zoom out to 24 km. At this scale it looks like a major landscape feature, easily the size of a plantation or a large lake - and it is - yet mining may not deserve the treatment it gets as an activity that rapes and pillages the landscape. For the impact of mines, even large ones like Cadia, is small compared to that of agriculture.

We need to get a wriggle on, otherwise our journey will take longer on the screen that it did in the car. So we type in the next destination, 'West Wyalong, NSW'. The distance is 174 km and our virtual flight will take us over thousands of paddocks and fields to flatter, drier country. There are still paddocks out here but they have less grass, carry less stock, and only really get going after good rain. Farming must still be practical in these conditions for every corner of the landscape has a field, but generally they are bigger and nearly all of them are cleared.

Even in the virtual world it looks hot out there, so we hop again, this time 226 km to 'Hay, NSW' where we find an intensive agricultural system. The farmers here are irrigators, using water from the Murrumbidgee River. The fields are small again. Aligned with the pipes they are impossibly angular and it seems that only the river and the areas prone to flooding escape the tractor. But this scene is deceptive. The intensive irrigation is an anomaly in this dry country. The next hop to 'Ivanhoe, NSW' is across 183 km of landscape without natural forests. There is no need to build that many fences for the paddocks are huge. They have to be for there is little for the animals to eat. Visually at least, this is the least modified of the landscapes we have seen around the stops on this virtual journey. It may not even be classed in the 58% of the world's land area that we have appropriated, yet there will be human influence on every pixel.

The penultimate hop is to 'Menindee, NSW', 185 km across this parched land. There are a few fields here because there is water from the lakes. The grey lateritic soils produced over the thousands of years of flooding are easy to see against the red sands. Dune slacks and crests are visible as stripes in the vegetation. The lakes are also here, although a zoom out to a height of 90 km is required to see them all. And there at the bottom left of the screen is Tandou Farm, the fields rectangular etchings on the lakebed.

This satellite view tells us that whilst we may have modified the lake to grow food and cotton, the impact on the landscape is no more than the mine at Cadia and far less than the tens of thousands of paddocks and arable fields to the east. No doubt there was a conservation impact of habitat loss for water dependent creatures when Tandou was created and, as with many a development, the debate as to the significance of this was muted. Yet, this view from an electronic bird says much about what might have been in the debate.

The big picture is rarely seen, yet this is what it takes to drive our collective.

Chapter 15

Story

Corridors of power

The Minister converses with his offsider.

"Brief, brief John, please."

"Yes Minister. We start at a farm that has an endangered wetland habitat. The farmer is third generation, although he's been around the traps before taking over from his old man. It's a good example of voluntary private reservation and management."

"OK, and our policy frame?"

"Revisions to the TSA now include this habitat type, so he has to preserve it anyway."

"My grab?"

"Private conservation provides essential support to the public reserve system."

"Excellent John, you're learning fast."

"Thank you Minister."

"When do we land?"

"In about half an hour, then it's a twenty minute drive with an interview with local radio as we go."

"No rest for the wicked."

Sylvia, John and Geoff work for the Ministry and discuss the Ministers visit.

"So what do you know about this new Minister?"

"Young, ambitious, articulate but full of the usual bull. More of the same."

"We certainly seem to get them."

"Better than the old buggers put out to pasture."

"True."

"So what are we supposed to tell him?"

"Very little according to Geoff. We let old Bruce do all the talking about his place, which is a given, and we help with any technical questions. Not that there will be any; oh, and we provide the guard of honour at the airport."

"Isn't he flying in to the Portman's strip."

"He is."

"Then 'airport' is a bit generous. 'Ladies and gentlemen, please fasten your seatbelts as we are about to land in a crudely leveled paddock.'"

"There will be a media frenzy."

"Do we get lunch?"

"We do."

"Tops."

Bruce and his wife Dorothy talk about the visit of the Minister to their farm.

"Bruce, is that you my dear?"

"Yes darling, it is me. Who else?"

"Did you mow the drive and open the gate?"

"Yes."

"And sweep the cobwebs from the hide."

"Yes."

"Can you hear me? Did you sweep the cobwebs?"

"Yes I swept the cobwebs. For goodness sake he's a junior Minister not the bloody Dalai Lama."

"Ah, there you are. Come sit, I've your tea."

"Honestly, what is it with all the fuss? We have had pollies here before. Department trots us out every time someone vaguely important pitches up from town. Sometimes I think we are the only farm Geoff steps foot on."

"Don't be churlish dear. It is important that we are nice to these people."

"We're more than nice. You've been cooking for a week. There are more cakes in there than the village bakery."

"There will be at least twenty people for lunch and I said I would make traditional farmhouse fare. So that is what I have done. You show them the property and I satisfy their stomachs."

"You'll certainly do that."

"So just you be on your best behaviour."

"Aren't I always? I'm just fed up with the dumb questions and the fake sympathy. When has a polly done anything useful as a result of visiting rural folk? What do we get for letting them stuff their faces with cake? Tell me, coz I'm kinda-"

"Now you of all people know that's not true. You, the one who is always first in the queue for the next subsidy, still complaining that it's not enough."

"Well it isn't."

"Ah, we haven't needed a cent of it and you know it."

"Don't be stupid woman, when is-"

"Just drink your tea. If you keep arguing you really will be too grumpy for our visitors."

The Minister and his offsider arrive.

"So this is the place?"

"Yes Minister."

"And the Department staff?"

"Sylvia is the blond, John is the fat guy and the short bloke in the hat is Geoff the regional manager."

"Very good. Any press?"

"I'm pretty sure the brunette over there is from the Ag review and the crew are local radio, the same mob that stood us up at the airport. The farmer at the gate is Bruce and his wife is Dorothy. Minister."

"Right, let's get to it."

The staffers make their comments.

"So that's him?"

"Yep, that's him. Quite cute if you ask me."

"Sylvia! How could you?"

"Quite easily actually, it has been a while since I ran my fingers through real hair."

"Yuk, too much information."

"Please people; some respect."

Bruce and Sylvia.

"Now dear are you sure you remembered everything?"

"Yes, I have remembered everything. I don't ask you if you put the cucumber in the sandwiches."

"No. But you are very forgetful these days."

"Well, if I have forgotten it's too late now, the bloke just got out of his car."

"My, he is a handsome one."

"In your dreams."

"Mmm, quite probably."

The Minister meets everyone.

"Geoff. G'day. It's a pleasure to meet you. Sylvia, John, everything humming along?"

"Yes Minister we like to think so."

"Looks like you have some good assets around the place, couple of fine yellowbox stands I noticed on the way in."

"Er, yes, we have been, er, working with the Landcare groups."

"Excellent. I am going to want that kind of initiative about the place. You had a problem with a glider population, localized sub-species I was told."

"Er yes, a few started appearing in gardens during the day and we lost quite a few to cats."

"So are we below MVP?"

"Erm, not sure that we know that Minister."

"Well it is obviously fragmented this little metatpopulation, so we either get to know its dynamics or we mitigate quick smart; no?"

"Er."

"Sylvia do we, perhaps, need some corridors?"

"Yeah, I mean, yes Minister. We proposed some that got up in the LEP and we could take a look at part of that today."

"Splendid. Now if you'll excuse me I must greet our hosts. G'day. Dorothy. Bruce. It is a pleasure. Thank you so much for inviting us down."

"G'day. You are welcome Minister."

"Peter please."

"Er Peter."

"Beaut property you have Bruce."

"Yeah, we're pretty proud of it. Been in the family a while."

"And you have a succession plan?"

"Yeah, my youngest son wants to take it on in a few years. Decided subdivision didn't really make much sense round here. Lucky my eldest chose the Army."

"I can see you need the acreage. What do you run, 500 head?"

"620."

"Mostly heifers?"

"Yeah, they go pretty well most years but it's been getting' harder with the drought."

"Didn't you get some good rains? Over 600 mm I was told."

"Oh yeah we had some showers, just that the land has a lot of catching up to do. Lots of catching up."

"I'm sure it will Bruce, it will. Dot, how are you?"

"I am very well Minister, thank you."

"I am told that your lunch makes it worth coming out all this way. Is that true?"

"No don't be silly, it's just what we would normally have."

"You are too modest, reputations are usually well deserved Dot. My spies also tell me that you are secretary to the local women's group."

"Yes I am, how do you find out all these things. I am..."

"Do not worry, we Ministers have our sources and they only tell me good things. It is very important to keep active in local politics. Do the ladies ever discuss the environment?"

"Er, no Minister. The men might let us get together for a tea and a gossip but not about the land."

"That's a pity Dot. I would like to know what you think. Maybe if you say I mentioned it, the ladies might agree to discuss it and send me their thoughts. I promise to consider them."

"That is very kind Minister, very kind. I will suggest it next Tuesday."

"Excuse me Minister."

"OK John. Sorry Dot, looks like I am being herded on. Looking forward to lunch."

"Minister, perhaps we should take a drive over to the wetland, there is a hide there and we can take Geoff with us."

"Splendid, lead on."

The Minister questions Geoff.

"So you see this is a very high quality reed bed habitat. It's obviously a little dry just now but it really does get going after a good shower."

"So Geoff, how do you know that it is good quality? All looks pretty dry and brown right now."

"Er yes, well, it's not great right now Minister, the last real wet was nine months ago, but we think that is pretty normal for these parts. When it was wet we had over 10,000 breeding pairs of knot, plus several thousand sandpiper out there on the mudflats, plus a whole heap of smaller birds nesting in the reeds."

"See this hide Peter, we put it up with an education grant. It's a ripper view."

"It is substantial Bruce. I can see you getting a whole class of school kids in here."

"Yeah, sometimes we do. They have to walk in though and the teachers get a bit antsy with all the mud an' all. Geoff here reckons we should have gone for a boardwalk but the money..."

"Sounds like something I should look into. Make a note John would you."

"Yes Minister."

"So Bruce, can we look at some of your paddocks?"

"Er, yeah, Peter. Sure you don't want to do a bit of spottin'?"

"Well thanks, but no thanks Bruce. Never been much of a birdo, although I hear you have records of Red-necked Phalarope which is a critter worth seeing in the breeding season, but I'm happy with the paddocks."

"Erm, sure, we can walk."

"Looks like you have a nasty weed problem Bruce, *Lipia* isn't it?"

"Er yeah, it's a problem keeping cover on this soil."

"Thought it would be pretty fertile, wetland derived, should be full of nutrients."

"Well yeah, it's pretty good, sure, but it needs the water before the grass really gets goin'."

"Geoff, what does the Department say should be done?"

"Minister, if you recall, your predecessor scrapped the invasive weeds program so we have no official recommendation for farmers in this district, but I have seen them try everything from ploughs to roundup mostly without any success. We think that burning is the way to go."

"On what fuel load?"

"Yes Minister, there rarely is enough ground cover to get a fire going."

"So fire is best but for that you need grass and grass is what you want to keep not to burn. This sounds like something we should be researching Geoff."

"Yes Minister."

"See Peter, down in these low paddocks ripping works for a while but after a couple of seasons the bloody stuff is back worse than ever. Like we are being strangled."

"Have you thought about the cell grazing option?"

"Ha! You are kidding. No way will that crazy idea work on this place Peter, no way mate."

"Why not? Seems like the kind of problem that approach was designed to fix, get the cows in intensively to rough the place up, the rest the place for long enough for the grass to get a hold."

"And you'll give me the cash for the fences?"

"You could get quite a bit done with your existing setup and some mobile electrics. Geoff you know all about this?"

"Minister?"

"Well Bruce here is a taxpayer and, along with a few others, he pays my salary, yours and for the funds we use to run our trials. I guess he might expect some sound advice that will help him decide after he's paid out all that income tax, not just what we think."

"Er-"

"Geoff, I am just kidding. Some things we will never know but must keep guessing with the help of some trial and error. Do you run any trials here Bruce? Seems like you have a good set up for it."

"Not these days Pete, although my old man did when I was a kid, he was into it, always having a barny with the extension officers and running around the paddocks with a notebook and his measuring tape. Seemed like a lot of extra work to me."

"So how do you know that chemicals will work on this *Lipia*?"

"Ah, I just spray it and see, paddock at a time when I can afford it."

"So you really don't know."

"Nope, you got me."

The staffers lick their wounds.

"What a smart-arse bastard."

"I think he is cute."

"Bollocks. Over inflated, up your arse politician pretending he's a scientist in his spare time. Ridiculous."

"...that a good looking bloke be a scientist and a politician."

"They even do courses in it, political science they call it."

"No need for you to get your fat gut involved."

"Just trying to help."

"Well he is cute, but he is obviously also clever enough to get the ever horizontal Geoff out of his rocking chair. I admire that."

"Next you'll be asking him out."

"No he is married; first thing I checked."

"Typical."

"So what's this fella said to make you mad?"

"Just how to do our job. I mean I don't go telling him how to make a speech or shake hands with the great unwashed."

"But what can he know?"

"Well everything apparently. Metapopulations, nutrient dynamics, stocking rates, bastard is a walking textbook."

"I like him, makes a pleasant change to have a boss who actually knows the trade."

"Ah, you just want him between your legs."

"Mmm, that would be nice too."

"Ssh, here he comes."

The Minister asks about the corridors.

"Sylvia, what can you tell me about these corridors?"

"Minister, if you would like to scan the valley ahead of us."

"Good looking country."

"So you see the patches of remnant veg, there in the middle distance to the left of the road."

"I see."

"They are mixed blackbutt patches, some in quite good condition and, well, the idea is to join them up by planting strips in the paddocks between them. We can plant more or less along the fence lines so it's only the edges of the paddocks we'll need to rehab."

"I see. What is in it for the farmers?"

"Er, the farmers Minister?"

"Sure, what is in it for them? Production benefits, pest control, offset credits or can he bank a biodiversity credit?"

"Er, none of those things are in the plan."

"So why would the farmers participate?"

"Well at the moment it is voluntary Minister, with some grant funds to help with the planting."

"And how do the farmers go about getting a grant?"

"There is an online application form."

"Just a sec Sylvia. Bruce, sorry Bruce excuse me mate."

"Yes Pete."

"Do you have internet access?"

"Ah, naw. Not into that porno nonsense. Do you need to-"

"No thanks, just curious. So I guess Sylvia, that Bruce is not a realistic sample, but there appear to be a few things that might get in the way of these corridors."

"Er, yes Minister. Ahm, Minister, can I speak openly?"

"Of course Sylvia, we are on the same team."

"I think that sometimes we get really good ideas and, well, the system seems to just shut them down with red tape. I mean, I thought incentives would be real important to get the farmers on board but not if they had to fill in an armful of paperwork that they all hate."

"I sympathise. It is a balancing act between getting the money out here to do the right thing and being seen to be squeaky clean in your processes back in town. So I agree there are impediments."

"Can you shift them for us?"

"Sylvia, that smile is wicked. You know very well that flattery will get you everywhere in politics, but not far in the bureaucracy. We probably need to be a bit smarter than that. Can you tell me exactly what these corridors will achieve?"

"Maybe Johnno should answer that, he studied them for his masters."

"Err, I, err Minister, it, I mean, yes, well."

"John, no worries, I am interested in the evidence we have that might convince Bruce here and his neighbours to put the effort into constructing corridors. What's in it for them?"

"Well, err, well not much actually. I mean, if we are being honest about it. I, well, I..."

"Go ahead John, I would like to know what you think."

"Well, I didn't find any evidence that corridors work Minister. I, I mean the literature is, well it's full of studies that say that they should. And, well, the theory seems to be logical enough, only I looked at birds and none of the species we are interested in keeping seem to use narrow strips of habitat at all. They, well, they like the security of patches, the larger the better. Most of the strips I looked at just get invaded by aggressive ferals, especially Indian Mynas. I mean I am probably completely wrong, err Minister."

"No John, I doubt that you are. Theory is only there to be tested after all; it is never a statement of truth. So, if I can summarise, we like the idea of corridors, but out here we have no evidence to show people that they actually work for the conservation objectives we have set."

"Er, yes Minister."

The Minister enjoys his lunch.

"Dot that was the most delicious pie, what did you call it again?"

"Beef and mushroom with spinach Minister, I mean Peter."

"Excellent stuff, you should start up a B&B, diversify the farm a bit, they'll be no shortage of customers once the word gets out."

"We thought about that a few years back only the council wouldn't give us planning permission."

"Really, why? Seems like an ideal activity for this fine place of yours. You must have half a dozen spare rooms."

"Yes we do. Oh, you must try this apple and cinnamon crumble. I can get you some ice cream to go with it."

"No Dot, thank you. I am under strict instructions from my wife, no more desserts. But do tell me why you were turned down."

"Oh, it is a pity about the crumble, it is a specialty of the house."

"Well, if you promise not to tell her, I'll have a small taste, but no ice cream, definitely no ice cream."

"There you are Min-, Peter. Enjoy."

"Thank you Dot."

"Well, you see, we applied and filled in all the forms, then ages later the council said it contravened their LEP."

"Really."

"Mate. I just don't get those buggers. I look after a patch of so called endangered habitat, and because I keep it, or at least my old man kept it, I can't have people over to take a closer look at it."

"Bruce is right Peter, they said it was because of our endangered habitat that we had to limit the number of people on the property. A B&B would bring in too many they said."

"That seems harsh."

"Can you do something Peter?"

"Well, very little I'm afraid Dot. Local Government matters come under planning, which is Joe Watermans portfolio. I can have a word with him, but cannot promise."

"No worries mate, good on ya for having a try."

More questions from the minister.

"So John, tell me about these LEPs. I thought we had planning sort out standards for them."

"They did Minister, unfortunately the regulations have fallen behind the legislation somewhat."

"Meaning."

"Meaning that several councils pushed through their LEP revisions before the new regs took effect. Just about all of them had a significant green lobby."

"That's a fair few."

"About thirty Minister, and over half of them changed their LEPs in a hurry. Bruce happens to be just inside the boundary of one of them; the greenest as it happens."

"Excuse me Minister, my name is Suzie Walker, I am deputy editor of the Ag Times. Do you mind if I ask a few questions?"

"On the record Suzie I hope."

"Yes Minister, of course. I am sure you have something interesting to say to our readers. So what brings you out so far?"

"Suzie, I wanted to see this region for myself. As you know I was at health and toured almost all the urban and provincial hospitals during my tenure. In my new job I can't get to every farmer, but I want to see as many as I can like Bruce here, to try and understand their challenges."

"So what have you seen?"

"Well Bruce and Dot have a property that is a terrific model. They run a successful commercial operation, yet they keep up the conservation values. Not only that but they bring people in to see it and that spreads the message."

"Surely the property is marginal given the lack of water and weed problems."

"Well Suzie, there are management challenges, but as you have seen, good partnerships with staff from my department are helping to fix some of them and it is these partnerships on private land that are so important to the success of the overall effort. They really help. Unfortunately we can't control the rain."

"How exactly do these partnerships work?"

"Well, you have been with us today. My staff have an innovative plan for linking remnant habitat across the landscape and we engage the farmers to contribute to that plan, putting on the ground committed conservation outcomes in a practical way. I cannot emphasis enough how important this private effort is to add value to our already comprehensive reserve system. As you know, just last week a further three reserves were gazetted adding another 10,000 hectares to the public estate. So private conservation really helps this effort and supports our reserve system."

"So you will repeal the TSA revisions?"

"Well we cannot be sure that the TSA revisions that parliament passed have not helped. It is very early days. I think we will need to look closely at the outcomes, especially on properties like this one, before we would consider any further changes. The revisions were about empowering private conservation effort, an effort that we must support if outcomes are to be achieved."

"So this is another election promise that the Government will not keep."

"No Suzie, I do not see it that way. The TSA was revised as part of a necessary modernization of legislation more than fifty years old, modernization the previous government failed to tackle. We have been bold and what we have seen today is a superb example of what that can do on the ground. So I believe that the revisions have been extremely effective. Our promise was to see more effective conservation on private lands. A promise no previous government dared to make and we have seen today that this government has delivered on that bold promise."

"Minister, thank you. Just one more question I know many of our readers would be keen to ask."

"Fire away."

"Why do you say one thing to Bruce, another to Geoff and a third to me?"

"No comment."

Chapter 16

Story

Grandfather

"Tsepho, my boy. You are well?"

"I am well my Kgosi." Tsepho replied bowing with his hands held together.

"Sit."

Tsepho sat cross-legged on a log beside his grandfather. The shade of an old Acacia tree framed polka dots of light on the bare ground in front of them.

Men had always sat under this tree and even though termites gave each log a finite life, logs had heard the conversations of the generations.

"You are well my chief?"

"I am old now. My teeth are gone, but I still see and I still sit."

The Kgosi sat on a traditional wooden stool leaning back against the bole of the tree, the lines on his face weathered to blend with the fissures in the bark behind his head. In his hand he held a wooden pipe that had seen as many moons as the logs, although Tsepho had never seen his Grandfather smoke it.

"Tell me," said the old man as he turned his head and raised both eyebrows to examine the lean, muscular youth, "tell me of the things you have seen. Tell me what wisdom you have found. But first, tell me how many rains have passed since you last sat with me."

"I am sorry my Kgosi, it is five. It is five rains since I last sat with you." The old man nodded slowly and tapped the pipe gently on his leg. Tsepho paused, his eyes searching aimlessly in the sand around his feet. "Kgosi, I have been working at Loluthi. At that place they are learning how to grow crops. They have what the big men call a Research Center and they gave me a job as an assistant. I went there two rains ago, at the start of the dry season. I was excited because I would still learn. You know that I wanted to continue with my studies, but this was just as good for me. There are many clever men working there and I knew they would teach me many things."

The old man continued to tap his pipe in a mysterious rhythm that Tsepho remembered from his youth but was never quite able to reproduce himself. There was a murmur and Tsepho took the hint to continue. "So I began to work. They made me dig deep holes all over a field that had grown maize for many seasons. Narrow holes, like mole burrows. Then I had to hammer a metal tube into each hole. There were many holes. I counted a hundred. It was a big field, many times wider than my father's fields. They ploughed it often with a big tractor, so the soil was smooth on the surface but very hard underneath. Then, three times each day and sometimes in the night, I carried a bag around the field for one of the big men. Usually it was Dr Jones." Tsepho flashed a toothy grin. "The bag was heavy but I was strong. At each hole I had to hold the bag still until the numbers stopped flashing on a green screen. Then I would write down the number on a sheet. Dr Jones told me that the numbers were measurements of the moisture in the soil at the bottom of the hole."

Tsepho looked up at his grandfather with a squint. The old man said nothing, closed his eyes and continued to tap out the same rhythm. Tsepho took a deep breath and went on. "I took great care and wrote the numbers down neatly and in the right place. Dr Jones was always happy with what I had done. At the last hole he would always say 'good work Tsepho, that's another one in the bag'. I did not know if he meant the bag I was carrying."

The old man opened his left eye and stared into the distance; the rhythm unchanging on his thigh. Tsepho shifted a little uneasily on the log, trying not to make it look obvious. "You are restless" said the chief again raising his head. "There is something in this story that you do not wish your Kgosi to know?"

"No my Kgosi, I am pleased to tell you my story."

"Then continue." Again the wrinkled eyelid dropped slowly over an eye that was as alive as if they had seen the light for the first time.

"Every day, I would carry the bag. Usually three, sometimes many times. Mostly it was Dr. Jones who would tell me to carry the bag and he would always walk with me. He knew I was curious, but I did not want to be disrespectful and I waited. Eventually, after many days, Dr Jones asked if I wanted to know more and I said that I did. I began to ask him questions, first about the bag and the holes; then about other scientific things. He was pleased with my questions and, I think, tried to answer with truth."

"Did your questions bring you answers?" The Kgosi widened his lips slightly as he spoke.

"Dr. Jones gave many answers Kgosi. He said that the bag was heavy with equipment that would record the smallest difference in the water that the soil held tight; water that the plant sends roots to seek. He said that if we measured this water change during the season when the crop grows, we can know much more about how the plant will use water."

"Did I not tell you that water is life? I told you when you were so small that even your fathers goats looked over your head."

"Yes Kgosi, I remember."

"Then go on, this wisdom of water you already have."

Tsepho looked away from this mild rebuke. It was a privilege to sit and share the shade with the chief of his tribe, the vessel of his heritage, the store that held the wisdom of the ages, passed down and remembered. It was the place for knowledge of all customs and ways from farming to warfare, courage to compassion. Absent from his mind Tsepho uncrossed his legs and placed his hands in his lap then quickly crossed them again, once more a small boy afraid he had done something wrong. "So," he said softly, "so I kept asking and soon I saw what happened to the numbers from the bag. Dr. Jones asked me to help him work with the computer. A very clever machine for counting Dr Jones said. I think you will have seen one Kgosi. It is a grey box, the color of the tiny termite mounds and a screen that is a deeper blue than the midday sky.

I would read the numbers I had written and Dr Jones would push buttons and the same numbers appeared on the blue sky. After a few days there were many numbers, more than any man could count." Tsepho paused and glanced across at the chief.

"A man need only count his cattle Tsepho. If he loses count, then he has too many," the Kgosi said without opening his eyes.

"Yes Kgosi, I remember. But Dr. Jones said that the computer could use these many numbers to help us understand. He said that..." Tsepho hesitated, cast his eyes down and lowered his head.

"Speak freely my boy. There is little I have not heard before. You may speak of these things."

Tsepho nodded. "Yes Kgosi. Dr. Jones said that the computer played with the numbers. I did not really know what this meant. I thought of the games I played as a child with stones and wire, but after a very short time the numbers changed to a picture. Dr. Jones said it was a scientific picture that he called a graph and it showed how the amount of water changed from the top to bottom of the holes. He showed me. More water at the bottom than the top." Tsepho traced the path of the graph with his finger then remembered his grandfather still had his eyes closed. "Many days passed and I continued to help Dr. Jones. Each day after the numbers went into the computer he showed me the graph. It was the same as the day before, like the seeds of an Acacia, curled in the same way each time. Dr. Jones would get very excited if the shape was different. On one day I told him how different the curve was before he noticed it and he was very pleased."

The old mans lips widened again, "And did this give you wisdom?" he said.

Tsepho hesitated. "No Kgosi, I cannot, I cannot say that it did. I learned to see when the curves were different, but it did not give me wisdom."

"Even the antbear knows that the earth is easier to dig once his claws break through the skin; the hardness that the sun has sealed with its touch. These are things that any warrior would know; would feel even before he washed his spear." The old man opened his eyes again and stopped tapping his pipe. He lent forward from the support of the tree and sat upright with hands on knees as though he was about to rise.

Tsepho jumped to his feet. "Is there anything I can get for you my Kgosi?" he said quickly and bowed again with his hands clasped out in front of him.

"No my boy, there is nothing," the old man smiled, "save for more of your story."

"Yes my Kgosi." Tsepho sat down again, responding to a wave of the hand holding the pipe.

"I carried the bag each day. When it was cold I carried it, when the heat of summer came I carried. Then something happened before the rain came. Dr Jones went away.

'One day he said, 'Tomorrow I'm going on a visit. Dr Rajpai will look after you while I'm away.' He smiled at me but I could see his eyes were wary. I had learned to watch Dr. Jones' eyes as they told me more than his words, which were always happy.

'At sunrise the next day I came to the field as usual. It was time for the rains and already there were some clouds in the sky. I waited for a long time and the sun was high before Dr Rajpai arrived. He was an angry man Kgosi. His shoulders were round and his brow was furrowed like the ploughed field. He always wore the same grey shirt and dark pants, even on the hottest day." Tsepho stopped, his jaw clenched tight and the muscles across his shoulders tied into knots. In the time it took to describe this man Tsepho was no longer sitting in reverence next to his grandfather.

The old man spoke softly. "Tsepho, do you remember when we would walk together? When we were silent, placing our feet with care, yet we moved without haste or purpose. Do you remember? I would point my pipe at a bird, and you would look at it with your whole body, not just your eyes. You would try to feel creation in it." Tsepho nodded slowly, then bowed his head. The Kgosi continued. "This big man Rajpai has hurt your mind, a poison has entered." Tsepho kept his head bowed, yet his shoulders eased slightly. "Close your eyes and remember our walks. Remember a time when I pointed to a quail. That bird, proud he was of his space and the nest that cradled his egg. Remember that a rival came close by, walking as we had without purpose, save for the need to find seeds among the stubble. The protective bird confronted the other with anger and aggression. They fought briefly in a flurry of feathers, beaks and clawed feet. Yet it was over, that anger, over as soon as it started. Both birds shook their feathers and flapped their wings, throwing away all that was bad. Again they were calm. It would seem, Tsepho, that you have not learned to flap your wings."

The young man kept his head bowed.

"Try this for your chief. Do you see that cow in the distance?"

Tsepho looked up slowly, readjusted his eyes and followed the line of the Chief's pipe. "I see it Kgosi" Tsepho said quietly.

"Then call it to me."

"Call it?"

"Yes, call it to me. Release the grip on your soul. Shout to that cow from the very pit of your stomach."

So Tsepho shouted to the cow.

"No, my boy, that was not a shout. That noise was just air in your lungs, try again. Try to find the place in your stomach where the hurt lives."

"Arret cow" Tsepho belted, but again it was only air.

The Chief leant across and put his hand on his grandson's stomach, "this time" he said with a calm, serene voice.

"Arret cow, arret" Tsepho yelled in a pure explosion of sound that traveled and echoed across the wide valley to the cow barely visible on the far hill.

The cow stopped grazing, turned and began to walk towards the village. The Chief nodded and his lips widened again into a real smile, gently took back his hand, waited for some moments and then said, "So now that you are calm, you can tell me about this man."

Tsepho held his stomach with both hands as a wave passed through him that melted the knots in his shoulder and the fear in his heart. He looked across at the old man and smiled at his grandfather.

"Yes, I can tell you now." Tsepho said.

'At first Rajpai refused to greet me. Then I was forced to do all the work for this man. When he saw that I could do everything that Dr Jones had taught me, he just sat in the shade and pointed to the field. This I did not mind, because I could be far from him. I worked this way for several days but it was not enough. He began to abuse me. Always he would shout or insult me. I cannot repeat what he said, Kgosi. There was nothing I could do. I did think I should leave, even to spear this man for his lack of respect, but I wanted to learn. The feeling of learning was stronger than my anger, so I remained quiet."

"There are times to shout and times to be silent. A wise warrior knows when to do battle. This I have told you."

"I remembered your wisdom Kgosi and because of this I kept quiet. And this only made his anger deeper. He began insulting my mother, my father, my tribe; even you my Kgosi. A warrior does not sit quietly when all he holds dear are attacked."

The breeze that had been gentle quickened, a dust devil materialized at the edge of the village and came towards them, its spiral top reaching above the canopy of the trees, then bent away as though fearful of its intended path.

The chief nodded and began tapping his pipe again.

"I endured the insults for three more days." Tsepho continued. "Then on the next day, it was a Saturday, this man he arrived later than usual. He stumbled out of his truck and walked slowly to me. I saw the red eyes that came with his unsteady walk. I knew that he had been at the shabeen, but it was still morning. Again he abused me with words, shouted at me to get my black ass into the field. He came towards me and pushed me hard so that I fell." Tsepho held his hands together, a grip tight enough to whiten his brown knuckles. A glower came to his face that enhanced his bold eyebrows and sharpened his jaw. "I was surprised by his attack Kgosi. I was already very angry at his words, but I felt ashamed that he had pushed me to the dirt. I wanted to strike him, to show him, to show him that I am proud, to revenge… to be a warrior and kill him with my hands.

"I jumped up and was ready to strike when another hand gripped me. It was Dr. Jones. He returned just as I was about to end my learning. My anger would have seen me dismissed, but he stopped me.

"He said to Rajpai, 'OK Ranjit, I will supervise today.' The drunk one was very angry and stared at my fist. There was fear in his face. 'Please' Dr. Jones said and Rajpai left.

"Dr Jones told me that he had seen everything and he apologized for what had happened."

"Did this give you wisdom?"

"Yes my Kgosi. I learned that some men can not control their demons... And that others are big men."

"That is wise..." the old man said, his voice trailing as though he is something more to add. Tsepho waited but there was no more.

"I worked only with Dr. Jones after that" Tsepho said after a long pause. "He continued to teach me. I learned how to use the computer and soon I was putting the numbers in myself. I even printed the graphs. Dr. Jones would come to see how I had done, and each time ask me what I had remembered and give me a new piece of learning. It was like being a boy and sitting with you my Kgosi. I felt proud that I was learning."

The chief smiled, closed his eyes and changed the rhythm to a slow, more lyrical beat.

"After the first rain, it was 17mm, we planted maize and sorghum. I told Dr. Jones that this rain was too little, that my father would know not to plant until it rained again. I told him that even the songololos had not woken from their winter sleep, but he said that it was okay. We could plant the seeds because the graph said there was enough water in the soil. So we planted, not in long rows but short rows and in odd patterns, sometimes maize, sometimes sorghum, sometimes both, all in the same field. Dr. Jones explained that it was an experiment. An experiment to see what effect maize and sorghum will have on the moisture in the soil.

"I remembered what you told me about how maize is always thirsty, so I said that maize would use more. Dr. Jones was pleased with this, but he said it was how the maize took drinks that was important. So I continued to measure from all the holes, reading the numbers from the heavy bag. Sometimes Dr. Jones was busy and he let me do everything myself. I was very happy on those days."

Tsepho stopped as a man he recognized as a friend of his father came towards the tree. The man's head was bowed and he held both hands together in front of him. He halted at the edge of the log circle and stood still. Tsepho nodded to acknowledge the man and he returned a glance of his own but said nothing. It was custom to let the chief speak first, even when it was more practical to announce your arrival.

After a minute that seemed to last longer, the chief opened his eyes. The man bowed his head even lower.

"Dumela Moses, you are well?" the Kgosi said.

"I am well my Kgosi. Blessings on you and your family." The man turned his head slightly to Tsepho.

"What brings you to your chief?"

"It is about logs my Kgosi. Termites have eaten the mopane poles in my families' grain store. It has fallen and soon the termites will eat..." He stopped realizing that he had gone on further than good manners permitted.

"So you need logs."

"Yes Kgosi."

"Has not Morapedi just cut logs to repair his mother's house?"

"Yes Kgosi."

The chief turned to Tsepho. "What should be done Tsepho?" He said then added "but I should tell you that while you were away, Morapedi gave Tolani a child."

Tsepho stiffened. Tolani was his friend, a beautiful girl that he might even have married one day, even though she was headstrong and had dreams of a modern life in the city. Tolani was with child by a man old enough to be her father. Conclusions lept at him.

"So Tsepho, what is the wise action for Moses?"

There was silence as Tsepho felt his thoughts chase each other through his body. Angry thoughts railed at the injustice, then fears that he might fail his Kgosi' and even sympathy for Moses; feelings that were harder to catch than mosquitoes at dusk. His chest heaved as he forced first one deep breath, then another.

"Kgosi, Morapedi should give his poles to Moses." Tsepho blurted with words that appeared from nowhere yet emerged clear and crisp. "And Moses should repair the house of Morapedi's mother with poles he must cut himself. Moses must do this as soon as his grain store is repaired."

Moses took a moment to realize what Tsepho had suggested, then looked across at him with indignation. Only the presence of the chief prevented him from venting. The Kgosi smiled again. “That is a solution Tsepho. I approve.” Turning to Moses, the chief added, “Moses, do you accept this wisdom?” Reluctantly Moses nodded but said nothing. He shuffled back with a few small steps, still with head bowed, then turned and walked towards the cow that was now near the edge of the village.

The Kgosi watched Moses leave then looked at Tsepho with his head a little higher than he had. Tsepho expected his grandfather to ask why, or even how, he came to his decision, instead the old man said, “Did the maize grow well?”

Tsepho was a little peeved and took a moment to remember where he was in his story. Once again the Kgosi closed his eyes and began to tap out an entirely different rhythm that Tsepho had not heard before.

“Erm, Kgosi, it was a good season in the beginning, with many rains. I had to record rainfall from four gauges, one in each corner of the field. Before the year ended there had been 256mm in more than 12 rain days. No more than ten days passed without rain.”

The chief lifted his eyelids to let Tsepho see the corner of his eye and the hint of a frown. “Sorry my Kgosi” Tsepho said, realizing his mistake. “It is easy to begin to speak like these research men. They have language of their own. The rains were good, there was water enough even for the thirsty maize and the seedlings grew well, much better than in my father's fields. Dr. Jones explained that in the past they fed the plants in this field and there would be food left over in the soil to feed the seedlings. All the time as it rained the numbers and the graphs changed shape because all the water was near to the surface.”

“That is why a truck gets stuck after rain.” The chief said as a matter of fact; then grinned.

Tsepho smiled himself and pressed on. “When the sun was overhead much of the day, it did not rain for many weeks. The soil dried and the songololos stopped walking. The numbers and graphs soon began to look like those before the rains and the maize plants sagged like the skin of the water dragon. The biggest ones became sick first, then the small ones. Some rains came for the rest of the season, but not enough to make the soil wet and only a few maize plants lived through the hot time and began to grow again. The sorghum was always better.”

"So your graphs told us we should grow sorghum in our fields."

"Em Kgosi, that is what they said."

"Is this not what we do? Do we not plant sorghum in many fields and maize in only a few?"

"Yes, Kgosi, this is what we do."

"Perhaps your heavy baggage and all its numbers will tell us what our ancestors learned and passed on to us so that we would know it."

"It is true my Kgosi. I learned much about how water moves in the soil, how to understand the numbers of those big men, even how to carry a bag with heavy machines." Tsepho stopped and his shoulders fell again. "But I did not learn how to grow more food for the children."

The old man opened his eyes and sat forward on the stool. "Tsepho, I am proud of you." He said. "You have brought honour to me, to your father, our village and our nation. You entered the white mans world and returned with dignity and wisdom. Nothing comes to you without awareness. Your choices are made with the help of the ancestors. One day you will be an honorable chief, my grandson.

'Now return to those big men and learn from them all you can.

'Come back and sit with me often. I am old, but I still have ears to listen."

Chapter 17

Coda

The awkward news is that we are a collective

In nature, *Homo sapiens* is just another prolific mammal species about a fifth of the way through an average mammalian 'lifetime'. We are just one of the almost unfathomable number of species that make up each and every habitat on earth. The black rat, house sparrow, and bracken fern are other well-known prolific species. Some piggyback on us and some have made it to ubiquity on their own. There have also been many prolific species in the past and others will evolve in the future. Everywhere common is not unique to us, nor is it something that we invented. We are part of the community of nature.

Individual humans come together to make societies, our collectives. These are as layered, complex and synergistic as the natural ones. Even though we have the same basic needs for our physiology and reproduction as any other large mammal, we have developed highly complex systems to deliver them. These systems appeared because our success gave us both the necessity and the time to invent. We reproduced in profusion and ended up living together in large, sometimes very large, groups. And yet, in the context of ecological and evolutionary time, the concept of urban is a complete novelty for the planet.

The economic and social system rules emerge because of us, in support of us, and they govern our use natural resources. As a result we no longer hunt and gather for our food, it arrives at the store on a truck. We no longer need to seek shelter in caves or under trees because we have systems for making resources easy to obtain and use. We also have the ability to adapt our environment to us. In this sense we have broken an evolutionary shackle.

In *Corridors of Power* the layers in our systems of resource use and governance were exposed, as were the kinds of people drawn to each of the layers. Diversity among us is both a strength and a challenge. The capacity to accommodate all views, ways of being and needs in a common system has left us perplexed more than once. Decision makers walk a tightrope unable to please all the people for even part of the time, so it attracts those with a strong sense of self-importance. This does not ensure sound compromise.

There is another human trait that is more fortunate. We all have a conscience; a sense of morals that puts a check on our behaviour. It is a sense that will have to become more acute and expand to include principles developed from ecological understanding. Thanks to his *Grandfather*, Tsepho was able to grasp that knowledge in itself was not enough for clear decision-making. There is something else, an added element as important as basil is to an Italian tomato sauce. Tsepho held it within him; we all do; yet he needed his grandfather to help him to release its power. Wisdom.

Part Four

The awkward news for Greenies

Chapter 18

In the end

Where is the problem?

This journey we have taken through the paradigm of sustainability, our connection to the earth and to each other provides one simple truth. This one truth is that **the environment does not have a problem**.

The environment is not an entity, it exists only as a place where a multitude of life forms and processes have come together and in which there resides a system of more-making that we call nature. It cannot have a problem with anything. It simply is.

In comparison to today, the environment of times past has variously been a place of much higher temperatures, global drought, wetness, weird configurations of landmasses, meteorite strikes, and volcanic eruptions of gigantic magnitude. There have been times with what now seem crazy levels of CO_2 in the atmosphere and times with next to no CO_2 in the atmosphere. And there has always been change in conditions; rapidly in some places and more sedately elsewhere. Sometimes in response to extra-terrestrial forces, but mostly as a result of long-term processes as the earth wobbles in its orbit, the tectonic plates slip along on the huge convection cells of molten rock in the mantle and the atmosphere and oceans respond in their own way. Nature, the biological part of the environment with its incessant more-making, has had a role too.

In the end, the environment is just a place where temperature, availability of water and wind cycle with the seasons and change on these longer rhythms. The only difference in change now from change past is that there is a species able to influence it, observe it, catalogue it, and attempt to understand it.

Nature does not have a problem either. Throughout geological time, the constant, persistent thing has been life. As far as we can discern, more-making has been copious for 3.5 billion years, is ongoing, and will continue long after we expire as individuals or as a species. When the world was hot, cold or shrouded in dust, nature was there, evolving, responding, coping and always more-making. Nature has persisted in all the conditions of the past, however extreme by comparison with today. It has persisted variously processing carbon and making more. If a complex community of life can evolve around the energy of deep-sea sulphur vents, then a bit of extra atmospheric CO_2 and a touch of global warming will be a cinch. So the problem is not with nature.

The problem is not with change. Change is what creates. It promotes diversity, the thousands of different ways of success in the creation of more. Nature is not concerned that this species able to observe, *Homo sapiens*, is also the latest major agent of change. Life will continue. Indeed, mass extinction simply creates the evolutionary opportunity for new diversity to emerge - over 99% of the species that have ever been are now extinct. It has happened many times in the past. Paleontologists have labeled them extinction events; short periods where many hundreds of thousands of species have been unable to cope with a change in the environment and died out. Yet nature persists and new organisms evolve to use the resources that become available.

The convention is that we are in, and through our alteration of habitat, pollution and resource appropriation we are the cause of, the sixth mass extinction event. The previous one was 65 million years ago, before Australia separated from Pangaea, at the Cretaceous-Paleogene transition when a sixth of all Families, half of the genera and three-quarters of the species went extinct. This was the event that ended the reign of dinosaurs and opened the way for mammals and birds to become the dominant land vertebrates. Even if the current one, caused perhaps because of a jump in atmospheric CO_2 concentrations on top of human land clearing, has a similar impact, life will persist. But it will not be with the current compliment of species.

The problem, this doubtful or difficult matter requiring a solution, is not with the environment, nor is it with nature or with the way evolutionary processes operate. The problem is not with or because of change. Despite our preference for stability, life exists in such a profusion of different types because conditions change. So where is the problem of the environment? Where is the great challenge of our time, this inconvenient truth?

The problem is **for us**.

It is also a problem **with us**.

The problem for us

What we have seen as the challenge, the demanding or difficult task, is to support a human population growing at 8,000 per hour. A population that not only grows in number but has disparity in standards of living creating a strong desire for betterment among those with less.

The problem is **for us** because we have to accept that 6.7 billion and counting has already appropriated close to half the primary production, altered most of landscape, modified the flow of freshwater and affected production in the oceans. The environment does not mind, nor does nature, but we will when our appropriation erodes the natural resource capacity and begins to change the environment in ways that put us at risk.[40] In short we will run out of resources and alter the processes that keep the environment as a healthy place for us. Some ecologists believe that in some parts of the world this has already happened. Some also think that, almost everywhere, we are close to critical breakpoints for resources such as water, nitrogen and soil fertility.

---

[40] Scientists have started to call the components of the environment that we find useful or require **ecosystem services** to distinguish them from natural resources, the components of the environment we harvest as products (goods)

The roots of our obsession with stability, sameness, keeping things constant are in an unconscious knowledge that when there is too much change in the environment it becomes a place less well suited to us. Yet we continue to clear land for agriculture, cut forest for timber and fish with abandon. Clearly the need for individual returns, be they subsistence or profit for those in business, is stronger than the feeling we might be eroding the resource. What the early urbanites saw in the land appeared limitless. The notion that trees, grass and water might be finite or that our actions might affect the capacity of nature to produce seemed absurd. And as they went about their business in town, they did not see the countryside change.

As our thought experiment told us, we live in the moment, and that moment is short and immediate. It is usual for us to need things now. We do not perceive the timeframes of the trees. So even though we crave stability, it is subservient to a much stronger need for gratification. And our short frame of reference obscures the long-term cycles of nature where change is what happens.

In the end our need to be users wins out. Now there are very many of us, all with the same need, the same fundamental drive to make more. This has given us environmental change; a change great enough to put at risk the very conditions we need to make more. This makes the environment a problem **for us**.

This message sounds familiar. Green groups and environmentalists have been saying something like this for a long time. Only they called it 'a problem with the environment' or 'environmental problems'. This gave us the false feeling that it was not our problem, but one for the environment. Hopefully we have put away that misnomer. The environment just is. It always has been and always will be 'just is'.

What has and will continue to change is that the environment may not deliver all the goods and services that we, that one species *Homo sapiens*, need for our more-making. When delivery fails, so do we.

The problem with us

How many of us are aware of the impact our everyday choices are having on anything other than ourselves or our immediate loved ones? We have to understand the consequences of our resource use, to touch the earth with an awareness that we currently lack.

It can be done.

It will be possible to combine the synergy of nature with our linear engineering skill to get much more out of the environment. It is likely that we can implement sustainability for most of our resource use. If we use our intelligence we can meet the increasing demand as the population grows and even out the disparity in affluence. There will be some losses, some things we will have to leave behind. Pragmatism will be a watchword and we will have to reconcile many differences of opinion.

Arguably this has already happened. Through innovation, the green revolution of the 1970's put back the doomsday predictions of some ecologists of the 1960's that our food supplies would run out. Resource rich people have donated to feed the starving in famine areas and to support those devastated by flood and tsunami. Innovation and altruism will happen again, only this time with more empathy for the synergy. It is in our power to solve the problem for us, should we choose to do so.

The problem is **with us** because we have to make the choices. We have to make the decisions that will determine our effect on the environment and it's affect on us. The problem really is one of choice.

There is no technological reason why we cannot feed all the people on earth, shelter them and allow them to make more. There is no knowledge we lack or detail that we cannot deduce. There is no reason why we cannot focus our problem solving to do this and at the same time save some of the space and resources for the other organisms on the planet.

The crux is can we choose? Are we capable of getting over ourselves? First by seeing that it is our choice and then making one? In this book we have seen why these are hairy questions.

We have also seen that in human beings this urge manifests in complex ways. We have a mind with extraordinary capacity and it is never idle. It works incessantly and when there is not a useful problem for it to solve it invents things of its own. These invented thoughts are confused and confounded by that innate desire for more-making and we find ourselves in fear for our survival, scared that we won't do it, paranoid that someone else will get there first or do it better. Modern humans are so disconnected from the reality of more-making that it is just imaginary. We have even invented non-biological ways of more-making; we call it making money. Yet the pressure of more-making still goes on in our heads. We fear lack, loss and death.

The question remains. Can we make the choice to moderate our change on the environment? It will mean that we must come out of our heads to touch the earth with our awareness. There is no doubt that we can do this. And even though humans continue to inflict pain and suffering our collective conscience is more aware than it was. However, being told to become aware does not create awareness. Similarly, rules cannot give us peace of mind and the clarity to solve the problem with us. We have to find this for ourselves, each one, one at a time. Before we can choose to moderate our resource use we must choose to be aware.

This is the problem **with us**. We have to learn that it is our choice. So far, few of us have chosen to be aware.

In the end

Think of someone you know who is around 15 years old. Chances are that your youngster lives in more worldly comfort than their great-grandfather would have thought possible. She is well fed, perhaps even a little overweight. She sleeps in a warm bed, goes to school and has any number of gadgets in her bedroom and in her bag. Most of the gizmos allow her to communicate on a whim with her many 'friends' no matter where they are or what they are doing.

She hears about nature and the environment in school and may even watch the occasional documentary on the Discovery channel, but mostly these things do not interest her. When the news comes on the television she switches to YouTube on her laptop.

In 2050 your youngster will be 57 years old. The United Nations medium projection has the global human population at around 9 billion at this time. This is roughly a third more people than we have today.[41]

When she reaches her life expectancy of 82 years in 2075 there will be anything from 4 billion to more than 30 billion people on earth (the low and high projections respectively) or maybe it will be 9 billion still. This is the medium projection. We are not sure. Despite our cleverness it is hard to project that far ahead.

We can be sure that she will have made some choices. Where to live, what to eat, whether to have children and how resources will have been used so that these needs can be met. She will have chosen how to satisfy her urge to make more - the same applies to men, they will have done this too – and she will have made a choice about the environment. Just as you and I make choices, so will she.

If you think she just followed her urges, stayed 'in her head', remained disconnected, carried on with business as usual, then read on.

If you think she followed her urges, but first got 'out of her head', reconnected with the earth, made choices, changed the way she did business, then skip to the last section.

[41] The upper projection for population growth will put the total in 2050 at somewhere around 12 billion and the lower at around 8 billion.

The end where we do not choose

It is 2075. There are more people that there have ever been. They say 15 billion, but we are not sure. Not everyone lives within the system.

There are wars everywhere because there are not enough resources to go around. Some people are fighting over water, some over land and many over food. The number of people living below subsistence has risen tenfold but subsistence now means a minimum of 3,000 calories per day, 3 litres of clean water, access to a sanitation cubicle and private sleeping quarter. Everywhere there have been losses. Mostly though the fighting is ritual.

People survive but they have lost dignity and a sense of purpose. This has been a huge challenge for the psychologists but the profession now attracts the best minds and has soared in prestige. Somehow though, there is enough goodwill to keep the collective from complete chaos.

There is ubiquitous clean power thanks to a clever boffin who cracked the fusion problem and this has helped. Without energy the desalination and pumping stations that water the intensive grain production sheds and deliver drinking water to mega cities would not work. Abundant power has also made manufacturing efficient and made us more mobile, especially in the virtual world.

Clever technicians have engineered away any clothing, sanitation and shelter needs. This is fortunate because without sterile living spaces we would succumb to any number of killer bugs. This is because the bacteria, and the fungi too, have been very happy to exploit all those human bodies. Generally though our bodies are healthy enough and spare parts are readily manufactured. You can pay to have the real thing cloned, but it is cheaper and easier to get bionic prosthetics.

There are still huge problems with any plants we have to grow. A quarter of us are scientists spending our time in a search for new strains of crop plant that can tolerate the conditions, ways of dealing with the waste from intensive production and working on the nirvana of cultivation on the moon. We lost the idea of ecosystems and their services decades ago as we now appropriate three quarters of the primary production. There are very few wild plants left. Nobody talks about biodiversity anymore.

Money still exists and, consequently, so does affluence. This also means that there are poor people too. No one has so far managed to break the 'we just have to have 10% unemployed' shackle that has been around for a century.

Most people just get along with life. They live for their VR goggles and are just satisfied that they have made more.

The end where we choose

It is 2075. There are fewer people now than in 1960, a tick under 3 billion.

There are wars here and there, not because there are not enough resources to go around, but there are sometimes some lingering misunderstandings about choice. Mostly though there seems to be no need to argue. The legal profession has dwindled on the lack of business.

People survive well as individuals and they make personal choices around their urge to make more. Curiously this has not resulted in population explosion but a contraction. The psychologists have put this down to the certainty of survival for both parent and child and the benefits of the collective to families.

There is ubiquitous clean power thanks to a clever boffin who cracked the fusion problem and this has helped. Energy was important for the desalination and pumping stations that water the intensive grain production sheds. This became vital after the decision to grow intensively in the harsher climates and soils, so leaving more of the naturally productive areas to supply ecosystem services. Grain farms in the Sahara, Arabian, Kalahari and Great Sandy deserts now provide half the worlds grain. There is some evidence to suggest that the Sahara is shrinking as the huge rainforest rehabilitation programs mature. Abundant power has also made manufacturing efficient and made us more mobile with more than two thirds of the population living part of their lives in the country.

Clever technicians have engineered away any clothing, sanitation and shelter needs. This has reduced the need for water and energy in houses, freeing the water harvested from roofs to go onto lush gardens. Gardening is easily the most popular pastime. People are healthy and spare parts are readily manufactured. The most efficient is through cloning from your own tissue but frequency of cancer and diabetes have declined since people chose exercise and to eat more fruit and vegetables grown without toxic pesticides.

There were some species losses, a legacy of the previous millennium, but the main habitat and ecosystem types have survived. No areas are protected anymore, as there is recognition that the whole system should be the focus of care. A quarter of the population spends some of their time and skill on the development or delivery of environmental management.

Money still exists and, consequently, so does affluence. This also means that there are poor people too. No one has so far managed to break the 'we just have to have 10% unemployed' shackle that has been around for a century. Employment for most is far less important than it was. The need for goods has diminished and with it the avarice for money.

Most people just get along with life. They live for their veggie patch and are satisfied with their choices.

Mark and Ashley would like to thank their loved ones, family and colleagues who have all influenced the content of this book.

If you have any questions or thoughts on *Awkward news for Greenies*, or experiences of your own that you would like to share, please contact us via our website.

www.thinktruth.com.au

Made in United States
North Haven, CT
04 November 2026

21763951R00117